Springer Desktop Editions in Chemistry

L. Brandsma, S. F. Vasilevsky, H. D. Verkruijsse
Application of Transition Metal Catalysts in Organic Synthesis
ISBN 3-540-65550-6

H. Driguez, J. Thiem (Eds.)
Glycoscience, Synthesis of Oligosaccharides and Glycoconjugates
ISBN 3-540-65557-3

H. Driguez, J. Thiem (Eds.)
Glycoscience, Synthesis of Substrate Analogs and Mimetics
ISBN 3-540-65546-8

H. A. O. Hill, P. J. Sadler, A. J. Thomson (Eds.)
Metal Sites in Proteins and Models, Iron Centres
ISBN 3-540-65552-2

H. A. O. Hill, P. J. Sadler, A. J. Thomson (Eds.)
Metal Sites in Proteins and Models, Phosphatases, Lewis Acids and Vanadium
ISBN 3-540-65553-0

H. A. O. Hill, P. J. Sadler, A. J. Thomson (Eds.)
Metal Sites in Proteins and Models, Redox Centres
ISBN 3-540-65556-5

A. Manz, H. Becker (Eds.)
Microsystem Technology in Chemistry and Life Sciences
ISBN 3-540-65555-7

P. Metz (Ed.)
Stereoselective Heterocyclic Synthesis
ISBN 3-540-65554-9

H. Pasch, B. Trathnigg
HPLC of Polymers
ISBN 3-540-65551-4

T. Scheper (Ed.)
New Enzymes for Organic Synthesis, Screening, Supply and Engineering
ISBN 3-540-65549-2

Springer-Verlag Berlin Heidelberg GmbH

P. Metz (Ed.)

Stereoselective Heterocyclic Synthesis

 Springer

Professor Dr. Peter Metz

Institut für Organische Chemie
Technische Universität Dresden
Mommsenstr. 13
D-01062 Dresden, Germany
E-mail: metz@coch01.chm.tu-dresden.de

Description of the Series

The Springer Desktop Editions in Chemistry is a paperback series that offers selected thematic volumes from Springer chemistry series to graduate students and individual scientists in industry and academia at very affordable prices. Each volume presents an area of high current interest to a broad non-specialist audience, starting at the graduate student level.

Formerly published as hardcover edition in the review series
Topics in Current Chemistry (Vol. 190)

Cataloging-in-Publication Data applied for

ISBN 978-3-540-65554-1   ISBN 978-3-540-68447-3 (eBook)
DOI 10.1007/978-3-540-68447-3

Cover: design & production, Heidelberg
Typesetting: Fotosatz-Service Köhler OHG, Würzburg
SPIN: 10711920   02/3020 - 5 4 3 2 1 0

Preface

Heterocycles play a central role in organic synthesis. Above all due to the interesting biological activities associated with a large number of these structurally diverse compounds, many heterocycles have been and will be challenging targets for total synthesis. Moreover, even if the final goal of a synthesis is not heterocyclic, at least a central intermediate or a key reagent used along the synthetic sequence most surely will be. This holds especially true if stereoselectivity is an important issue, as modern heterocyclic chemistry provides the synthetic organic chemist with an excellent arsenal of methods and strategies for the stereocontrolled construction and elaboration (including the cleavage) of heterocycles. Recent years have witnessed exciting new findings in this field, and it is the aim of "Stereoselective Heterocyclic Synthesis" to present a selection of these novel developments.

As the guest editor I am very glad that leading researches in this area have contributed highly inspiring accounts with up-to-date coverage to this compilation. The book comprises chapters on *"Using Ring-Opening Reactions of Oxabicyclic Compounds as a Strategy in Organic Synthesis"* by *P. Chiu* and *M. Lautens* focussing on the preparation and the synthetic utility of the versatile title compounds, *"The Nucleophilic Addition/Ring Closure (NARC) Sequence for the Stereocontrolled Synthesis of Heterocycles"* a powerful tactical combination discussed by *P. Perlmutter*, *"Chiral Acetylenic Sulfoxides and Related Compounds in Organic Synthesis"* by *A. W. M. Lee* and *W. H. Chan* emphasizing the use of sulfur-activated acetylenic and vinyl units for the efficient preparation of heterocycles, and *"N-Sulfonyl Imines - Useful Synthons in Stereoselective Organic Synthesis"* by *S. M. Weinreb* giving a comprehensive review on the chemistry of these valuable electron-deficient compounds.

I hope that the articles collected will not only serve experts in the field but will also attract the interest of scientists not yet familiar with this fascinating research topic.

Dresden, March 1997   Peter Metz

Topics in Current Chemistry
Now Also Available Electronically

For all customers with a standing order for **Topics in Current Chemistry** we offer the electronic form via LINK **free of charge.** You will receive a password for free access to the full articles.
Please register at: **http://link.springer.de/series/tcc/reg_form.htm**

If you do not have a standing order you can nevertheless browse through the table of contents of the volumes and the abstracts of each article at:
http://link.springer.de/series/tcc

There you will also find information about the
- Editorial Board
- Aims and Scope
- Instructions for Authors.

Contents

Using Ring-Opening Reactions of Oxabicyclic Compounds as a Strategy in Organic Synthesis

Pauline Chiu[1] and Mark Lautens[2]

[1] Department of Chemistry, University of Hong Kong, Pokfulam Road, Hong Kong
E-mail: pchiu@hkusua.hku.hk
[2] Department of Chemistry, University of Toronto, Toronto, Canada M5S 1A1
E-mail: mlautens@alchemy.chem.utoronto.ca

This chapter discusses the various methods for the preparation of oxabicyclic compounds, with an emphasis on the stereo- and enantioselective synthesis of these substances. Methods to desymmetrize *meso* oxabicyclic compounds are also presented. The utility of these substrates for organic synthesis is demonstrated by the many strategies available for ring-opening the bicyclic compounds to yield cyclic and acyclic structures. Examples demonstrating how this strategy has been incorporated into the efficient syntheses of many natural products are presented.

Keywords. Cycloaddition, oxabicyclic, ring opening, stereocontrol, natural products

Table of Contents

Topics in Current Chemistry, Vol. 190
© Springer Verlag Berlin Heidelberg 1997

List of Abbreviations

9-BBN	9-borabicyclononane
BINAP	2,2′-bis(diphenylphosphino)-1,1′-binaphthyl
Bz	benzoyl
COD	cyclooctadiene
dba	dibenzylideneacetone
DDQ	dichlorodicyanoquinone
DIBAL-Cl	diisobutylaluminum chloride
DIBAL-H	diisobutylaluminum hydride
DMAD	dimethylacetylene dicarboxylate
DME	dimethoxyethane
dppb	1,4-bis(diphenylphosphino)butane
HMPA	hexamethylphosphoramide
LAH	lithium aluminum hydride
LDA	lithium diisopropylamide
LHMDS	lithium hexamethyldisilazide
LiDBB	lithium di-*tert*-butylbiphenylide
mCPBA	*meta*-chloroperoxybenzoic acid
MS	molecular sieves
PMB	*p*-methoxybenzyl
PMP	*p*-methoxyphenyl
PPTS	pyridinium *p*-toluenesulfonate
pyr	pyridine
Red-Al	sodium bis(2-methoxyethoxy)aluminum hydride
TBDMS	*tert*-butyldimethylsilyl
Tf	trifluoromethylsulfonyl
THP	tetrahydropyran
TMP	2,2,6,6-tetramethylpiperidine
TMS	trimethylsilyl
Tr	trityl
Ts	toluenesulfonyl

1
Introduction

The use of rigid polycyclic templates to influence the stereoselectivity of functional group introduction or interconversion, followed by cleavage reactions to form simpler rings or acyclic chains, has been a common strategy in the syntheses of many important natural products. The analogous exploitation in the context of oxabicyclic templates has also increased as the repertoire of reactions for the synthesis and ring opening of these compounds has grown.

Interest in ring cleaving reactions of oxabicyclic compounds experienced significant growth in the late seventies as a consequence of the development of new methods to assemble oxabicyclo[3.2.1] compounds. Ring opening of oxabicyclo[2.2.1] substrates also underwent a renaissance in concert with

1

x, y≥1

Fig. 1

improvements in the Diels-Alder reaction of furans. The studies of these systems have made oxabicyclic substrates attractive starting materials in organic synthesis. Both monocyclic as well as acyclic compounds have been prepared using ring opening reactions.

In this review, methods for the construction of oxabicyclic substrates of general structure **1**, Fig. 1, are described as well as ring opening reactions which are applicable to the synthesis of natural products.

2
Preparation of Oxabicyclic Substrates

The aim of this section is to give an overview of the general approaches that have been employed. A comprehensive review of all of the methods used to date for the synthesis of oxabicyclic compounds is beyond the scope of the review. Instead, a focus on the preparation of those oxabicyclic systems for which ring opening reactions have been developed is described. The more recent achievements in this area will be emphasized.

2.1
Cycloadditions with Furan Derivatives

2.1.1
[4+2] Cycloadditions with Dienophiles

The Diels-Alder reaction, employing furan and substituted furans, has been the most widely investigated strategy to construct the oxabicyclo[2.2.1]heptene framework. Furan is not very reactive as a diene due to the loss of aromaticity which accompanies the cycloaddition. Among the solutions investigated to date to improve the reaction are catalysis using Lewis acids [1], metal salts and complexes [2], Cu^{2+} [3], silica gel [3], zeolites [4, 5], ultrasound [6], centrifugation [7], and high-pressure techniques [8]. Owing to the abundance of the instances of furan [4+2] cycloadditions in the literature, and the reviews that have already appeared [9, 10], highlights of this reaction in the cases where extensive work has been published on the subsequent ring opening will be outlined.

Vogel has developed 7-oxanorborn-5-en-2-one **2** and its derivatives as versatile alternative synthons for sugar chirons; hence these substrates have been coined "naked sugars." Several reviews summarizing this work have appeared [11]. The key cycloaddition in the synthesis is a Diels-Alder reaction between furan and 1-cyanovinylacetate **3a** catalyzed by ZnI_2, Eq. 1. The initial mixture of *exo* and *endo* cyanohydrins is equilibrated with base, resolved using brucine and acetylated to provide a 7:93 mixture of *exo/endo* cycloadducts. Successive recrystallizations of the initial crop afforded a 14% yield of the *endo* isomer (+)-**4a** of >99% ee [12].

Enantiomerically enriched products can also be obtained by employing a dienophile bearing a chiral controller group [13]. For example, the use of the camphanate ester derivative (S)-**3b** (also available in the (R) form) in the cycloaddition with furan gave a 29% yield of diastereomer **4b** after purification, along with other *endo* and *exo* isomers, Eq. 2. Saponification afforded the chiral ketone (+)-**2**. Reactions of **4b** and **2** have been reported to occur with high regio- and stereocontrol (vide infra).

"Naked sugars of the second generation" **5** have since been developed based on the cycloaddition between 2,4-dimethylfuran and **3b**, Eq. 3 [14]. Because both enantiomers of the naked sugars are available in large quantities from relatively inexpensive starting materials, they represent an important family of chiral substrates available to the synthetic chemist.

Koizumi observed a diastereoselective cycloaddition between furan and chiral vinyl sulfoxides **6** [15]. While the analogous *p*-tolylsulfinyl acrylates were completely unreactive in this reaction, the 2-pyridyl (Py) substituent significantly enhanced the reactivity of the vinyl sulfoxide toward cycloaddition. Separation of the diastereomeric sulfoxides was much easier when menthyl (Menth) acrylates were used. The cycloaddition of chiral sulfinyl menthyl acrylates 6 with furan proved to be highly diastereoselective, as illustrated in Eq. 4. The cycloaddition of 6 and the more reactive 3,4-dibenzyloxyfuran occurred at –20 °C to give products with comparable diastereomeric ratios (dr's) but in higher yield [16].

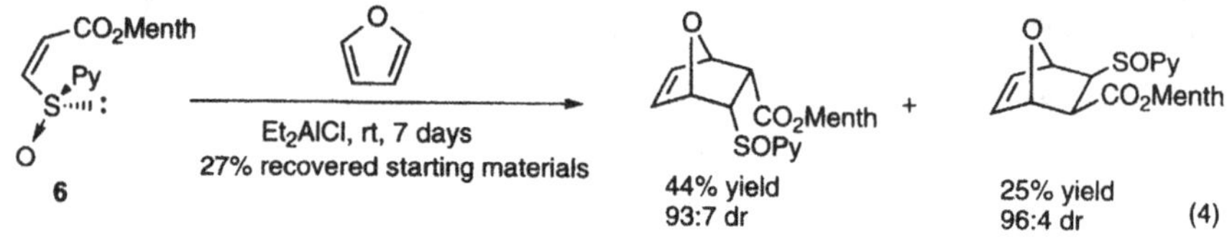

An extremely efficient and high yielding catalytic asymmetric Diels-Alder reaction of furan was reported by Corey [17]. In the presence of 10 mol% of oxazaborolidinone 7, Fig. 2, the cycloaddition between furan and 2-bromoacrolein proceeded to give the *exo* oxanorbornene derivative 8 in excellent yield and 92% ee, Eq. 5. Compound 8 can be efficiently converted to oxanorbornenone (+)-2, which is otherwise obtained by the Vogel methodology.

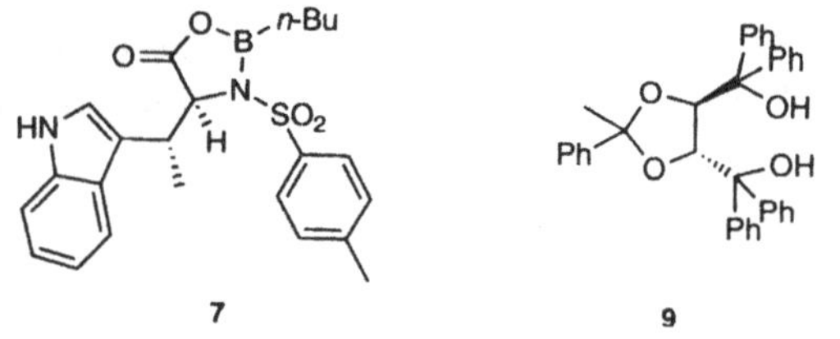

Fig. 2

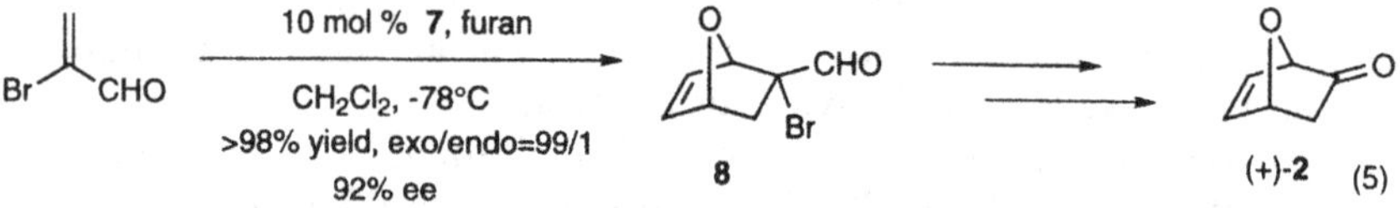

A methylsulfido substituent on the furan significantly enhances its reactivity toward dienophiles. Therefore, cycloadditions of 3-methylthiofuran proceeds even with monoactivated olefins to give predominantly the *endo* adducts, Eq. 6 [18]. Moreover, chiral titanium catalysts generated in situ from $(i\text{-PrO})_2\text{TiCl}_2$ and tartrate derivative 9, Fig. 2, induce cycloaddition with 3-acryloyl-1,3-oxazolidin-2-one with good enantioselectivity and in excellent yield, Eq. 7.

1 wt % hydroquinone

Et$_2$O or neat, 2-3 days

R= COOMe, 74% yield   85 : 15
R= COMe, 89% yield   90 : 10
R= CHO, 69% yield   62 : 38   (6)
R= CN, 78% yield   66 : 34

10 mol % **9**
TiCl$_2$(O-iPr)$_2$

MS 4 A,
PhMe- petrol ether
-10 to -5 °C, 2.5 h

R= H, 97% yield   87% ee   85 : 15
R= COOMe, 99% yield   86% ee   78 : 22   (7)

The efficiency of the intramolecular Diels-Alder reactions of furan has been described in several reviews, including an excellent treatise by Lipshutz. Steric factors, rather than electronic or solvent effects, appear to have the greatest influence on the outcome of the cycloaddition [1, 19, 20]. Electronically-disfavored cycloadditions can be brought about by creative functional group modifications. Thus, an electron-deficient furan, such as one bearing an α-keto group, can be masked and induced to undergo cycloaddition, as shown in Eq. 8 [21].

1. BuNEt$_3$$^+F^-$

2. dilute NaOH, 84% yield

77%   88:12   (8)

Metz has developed a highly diastereoselective intramolecular Diels-Alder reaction of furans with vinyl sulfonates [22]. When hydroxyfuran **10a** was esterified with vinylsulfonic acid chloride, the intermediate sulfonate spontaneously underwent cycloaddition to give sultone **11a**, Eq. 9. In the same manner, (–)-**11b** was obtained from (R)-**10b** which was derived from L-valine.

Another diastereoselective intramolecular Diels-Alder reaction of furan was studied by Keay wherein the methyl group in the tether of (–)-**12** directed the facial selectivity of the cycloaddition. Equilibrating conditions using a catalytic amount of Lewis acid gave the tricyclic enone (–)-**13**, Eq. 10 [23].

10a R= Me
(R)-10b R= i-Pr

11a R= Me, 90%
(+)-11b R= i-Pr, 87%

$$(9)$$

cat MeAlCl$_2$

CH$_2$Cl$_2$, -78°C

92% yield

(–)-12

(–)-13   89:11

$$(10)$$

2.1.2
[4+3] Cycloadditions with Oxyallyl Cations

The furan nucleus undergoes cycloadditions with oxyallyl cations to produce compounds with the oxabicyclo[3.2.1]octene skeleton. Various research groups have found new ways of generating the oxyallyl cation and have also defined the types of substituted furans which undergo reaction. Reviews on this reaction have covered the literature up to 1987 [24–26]. The mechanism of the cycloaddition has been discussed in detail by Hoffmann [26].

A more recent development in the generation of oxyallyl cations from polybromoketones has been the use of diethylzinc [27]. This procedure is convenient and amenable for the large-scale syntheses of oxabicyclic compounds. In addition, the combination of cerium (III) chloride and tin (II) chloride has been very effective in inducing the [4+3] cycloaddition between furan and 2,4-dibromopentan-3-one [28]. Sonication has also been observed to improve yields in cycloadditions promoted by zinc-copper couple [29].

The cycloadditions of several unusual oxyallyl cations are briefly outlined in Table 1. Tricyclic oxa-bridged substrates can be readily assembled from cyclic oxyallyl cations derived from monohalogenated cyclic ketones **14a** and LiClO$_4$/Et$_3$N (entry 1) [30]. Dihalogenated cyclic ketones **14b** can also serve as cyclic oxyallyl cation precursors when treated with diiron nonacarbonyl [31], or zinc-copper couple [32]. Cycloadditions of this type have been successful in producing oxatricyclic compounds where n = 2, 3, 4, 5, 9, although some adducts are mixtures of *cis* and *trans* isomers.

Oxyallyl cations bearing oxygen substituents have been synthesized from **15** and catalytic TMSOTf (entry 2) [33], from **16** and LiClO$_4$/Et$_3$N (entry 3) [34], and from the reaction of pyruvaldehyde directly with SnCl$_4$ (entry 4) [35]. Oxyallyl cations from **17** undergo cycloaddition with furan to give nitrile-substituted oxabicyclic compounds (entry 5) [34b].

Conjugated and homoconjugated oxyallyl cations from **18, 19a**, and **19b** have also been trapped with furan to give cycloadducts (entries 6, 7) [36].

Table 1. Synthesis of [4+3] Cycloadducts from Complex Oxyallyl Cations

Entry	Oxyallyl Cation Precursor	Furan Derivative	Reaction Conditions	Oxabicyclic Product (Yield)	Ref
1	**14a** X_1=H, X_2=Cl **14b** X_1= X_2=Br		A= **14a**, Et_3N $3M$ $LiClO_4$ in ether B= **14b**, $Fe_2(CO)_9$, Δ C= **14b**, Zn-Cu	n= 2, A(64%) n= 3, A(81%), B (35%) n= 4, A(11%), B (54%) n= 5, B (37%) n= 9, A(56%), B (52%)	[30-32]
2	**15**	R= R'= H R= H, R'=Me R= R'= Me	TMSOTf, $EtNO_2$, -78°C	R= R'= H (67%) R= H, R'=Me (54%) R= R'= Me (78%)	[33]
3	**16**		$3M$ $LiClO_4$ in ether Et_3N	(44%)	[34]
4			$SnCl_4$, CH_2Cl_2, -78°C	(43%)	[35]
5	**17**		1. Ag_2O 2. Zn-Cu, MeOH	(75%)	[34b]
6	**18**		$SnCl_4$, CH_2Cl_2, -78°C	(36%)	[36]
7	**19a** X=O, R= H **19b** X=CH_2, R=CO_2Et		TMSOTf, -50°C or $TiCl_4$, 0°C	R' = OH 12% R' = CH($CO_2Et)_2$ 15%	[36]

$$(11)$$

The *endo* peroxide **20**, a precursor of cyclopropanone **21**, yields [4+3] cyclo-adducts with furan when heated, Eq. 11 [37].

Although furans tethered to α,α'-dibromoketones undergo intramolecular [4+3] cycloadditions with diiron nonacarbonyl and with lithium perch-lorate/triethylamine, the modest yields of adducts that were obtained motivated additional studies [38, 39]. Harmata has extensively examined intramolecular [4+3] cycloadditions using various oxyallyl cation precursors and also investiga-ted the mechanism of the reaction [40–43]. Under Lewis acidic conditions, alk-oxy allylic sulfone **22** generates an allylic cation for cycloaddition, Eq. 12. Under the same reaction conditions, vinyl thioether **23** was found to generate an alk-oxyvinyl thionium intermediate that undergoes cycloaddition with the tethered furan, Eq. 13. Most recently, Harmata has shown that appropriately substituted allylic alcohols bearing a tethered furan generate vinylthionium ions in the presence of triflic anhydride which react to give [4+3] cycloadducts [42b]. Harmata also found that the alkoxyallylic sulfoxide **24** undergoes a Pummerer rearrangement to yield the thionium intermediate, which undergoes an intra-molecular cycloaddition in high yield, Eq. 14.

$$(12)$$

$$(13)$$

$$(14)$$

The construction of tetracyclic substrate 25 has been achieved by the intramolecular cycloaddition of a furan tethered ($n=1$) to a cycloalkanone using conditions related to those developed by Föhlisch, Eq. 15 [42, 43]. As the ring size of the oxyallyl cation increased, products arising from cycloaddition via the less strained *exo* transition state predominated. Cycloadducts with $n=6, 8$ have also been successfully prepared as a mixture of stereo isomers [43b].

$$\text{(15)}$$

Photogenerated oxyallyl cations undergo intramolecular cycloadditions when tethered to a furan; an example is illustrated in Eq. 16 [44]. The cyclohexadienone precursors required substituents which provide the right electronic characteristics for the photoconversion to the oxyallyl zwitterions, therefore limiting the versatility of this reaction.

$$\text{(16)}$$

Recently, Lautens, Aspiotis and Colucci extended the [4+3] cycloaddition methodology to include the diastereoselective intermolecular cycloaddition between an oxyallyl cation and a chiral furan [45]. The best results were obtained employing furan 26 bearing a free hydroxyl group in the 2-position, reacting with excess 1,3-dibromopentanone in the presence of diethyl zinc. Under the optimized conditions, up to 80 % yield of the crystalline oxabicyclo[3.2.1]octene 27 was obtained with a diastereoselectivity of $\geq 19:1$. The other product was the minor diastereomer 28, Eq. 17.

A significant observation was the unusual stereochemistry of the methyl groups at C_2 and C_4 in 27 which were pseudoaxial rather than pseudoequatorial. Instead of the cycloaddition occurring via transition state A (Scheme 1), where

$$\text{(17)}$$

$$\geq 19:1$$

Scheme 1

the oxyallyl cation assumes the most stable W-configuration and the cycloaddition is in the compact mode, the predominance of diastereomer **27** is postulated to occur via transition state **B** in the extended mode due to simultaneous coordination of the oxyallyl oxygen and furan side-chain oxygen to zinc. This appears to be the first example of a [4+3] cycloaddition of furan that has proceeded predominantly via an extended transition state. Diastereomer **28** probably arose from the cycloaddition of the less stable sickle-configuration of the oxyallyl cation or via a stepwise process.

2.1.3
[4+3] Cycloadditions with Allylic Cations

In the presence of zinc chloride, [4+3] cycloadducts between the allylic cation formed from **29** and furan are obtained, Eq. 18. However, the major product of the reaction arises from electrophilic addition to furan [46].

$$\text{(18)}$$

Harmata has reported the formation of intramolecular cycloadducts derived from trimethylsilylmethyl allylic sulfones **30**, Eq. 19 [47]. Optimized reaction conditions involved the use of trimethylaluminum as the Lewis acid.

Furthermore, **30** also afforded the same cycloadducts under photolytic conditions, Eq. 20. While the reaction is not synthetically useful in its present form,

$$(19)$$

$$(20)$$

it appears to be the first example of the photogeneration of an allylic cation for [4+3] cycloaddition [48]. The modest yield is probably due to the simultaneous degradation of the product under photolysis.

2.1.4
[4+4] Cycloadditions with Pyrones

West has recently reported intramolecular [4+4] cycloadditions of furan and 2-pyrone under photolysis in aqueous solution [49]. The *exo* and *endo* adducts are obtained in varying ratios when **31** bearing a variety of substituents is photolyzed in an aqueous solution of LiCl, Eq. 21.

In substrate **32**, modest diastereoselectivity was observed as a result of the preexisting stereocenter in the tether, Eq. 22.

31a Z = O
31b Z = S
31c Z = C(CO$_2$Me)$_2$

Z = O, 66% yield   1.5:1
Z = S, 61% yield   2.4:1
Z = C(CO$_2$Me)$_2$, 49% yield   1:2.4

$$(21)$$

32

37% yield
α:β Me 5.2:1

29% yield
α:β Me 1.3:1

$$(22)$$

2.1.5
Cyclopropanation/Rearrangement of Furan Derivatives

Rhodium carbenoids react with furan derivatives to generate oxabicyclo[3.2.1] octadienes through the formation and rearrangement of divinyl cyclopropane intermediates. Therefore, treatment of 2,5-dimethylfuran with **33** leads to the *endo* adduct **34**, Eq. 23 [50].

Davies recently reported a highly diastereoselective version of this reaction by incorporating chiral auxiliaries within the carbenoid precursors [50c]. Thus, the rhodium-catalyzed reaction of 2-methylfuran with vinyldiazomethane **35** afforded cycloadduct **36** with 94% de, Eq. 24. The products can be obtained in >99% de after purification by flash chromatography.

$$(23)$$

$$(24)$$

Furthermore, an enantioselective cycloaddition was observed when the carbenoid was formed by the use of a chiral rhodium complex **37** derived from (*S*)-*N*-*para* (*tert*-butylbenzene)sulfonylprolinate (TBSP). For example, decomposition of **38** by **37** in the presence of furan generated **39** with 80% ee, along with a triene containing side-product in 15–20% yield, Eq. 25. However the ee dropped significantly when other vinyldiazo compounds were studied under analogous conditions.

$$(25)$$

2.2
[5+2] Cycloadditions of Pyrylium Betaines

Metastable, aromatic pyrylium species undergo cycloadditions with olefins to generate [3.2.1] oxabicyclic derivatives. This reaction has been reviewed by Sammes and by Katritzky [51–52].

Hendrickson pioneered the use of pyranulose acetates with the basic structure **40** as a precursor for oxidopyrylium ion **41**, Eq. 26 [53]. Thermolysis of **40** in the presence of unhindered, electron-deficient olefins and acetylenes led to the formation of oxabicyclo[3.2.1] compounds.

$$(26)$$

Subsequently, Sammes showed that while simple olefins were unreactive, strained olefins such as norbornadiene, and electron-rich olefins such as vinyl ethers also undergo cycloaddition with **41** [54]. In the latter case, the *endo* cyclo-adduct tends to predominate. Sammes also successfully induced the formation of **41** from **40** in the presence of catalytic base at ambient temperatures, thus allowing the cycloaddition to proceed under mild conditions, Eq. 27.

Intramolecular versions of this cycloaddition were observed starting from 2-substituted and 6-substituted oxypyrans **42** and **43**, Eqs. 28, 29 [55–57].

$$(27)$$

$$(28)$$

$$(29)$$

Wender extended the studies of the intramolecular cycloaddition by examining substituted oxidopyrylium intermediates with stereocenters in the tethers [58a]. Pyran **44** underwent smooth cycloaddition with complete stereoselectivity to give **45** due to the methyl group at C_{11} assuming an equatorial position in the chair-like conformation of the olefinic side-chain, Eq. 30. The stereocenter at C_{11} effectively controlled the stereochemistry at C_6, C_8, and C_9. The reaction proceeded with heating or at room temperature with a catalytic amount of base.

Williams and Lupi independently provided additional examples of intramolecular cycloadditions, Eqs. 31, 32 [59–60].

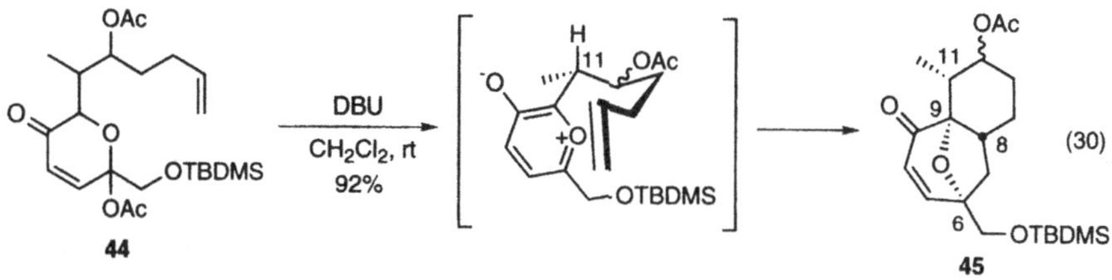

(30)

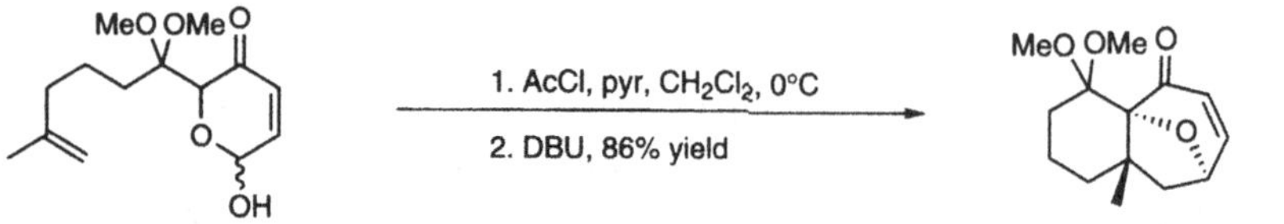

(31)

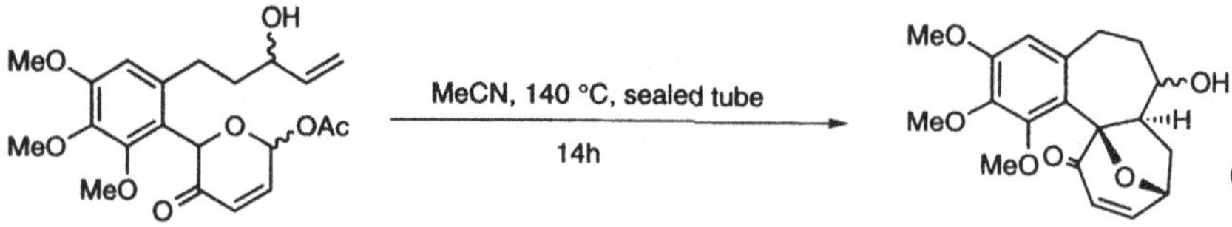

(32)

Garst investigated the intramolecular cyclization of substituted 4-pyrones derived from kojic acid, such as **46a**, with internal olefins (33) [58b]. The cycloaddition occurs under thermal and, in some cases, acid-catalyzed conditions. Substrate **46b** with a tether bearing a nitrogen also undergoes cycloaddition. In general, the substrates which undergo cycloaddition have tethers with three or four atoms and result in fused [5, 7] or [6, 7] ring systems, although product **47** with an intervening 7-membered ring has been obtained with an aromatic substrate, Eq. 34.

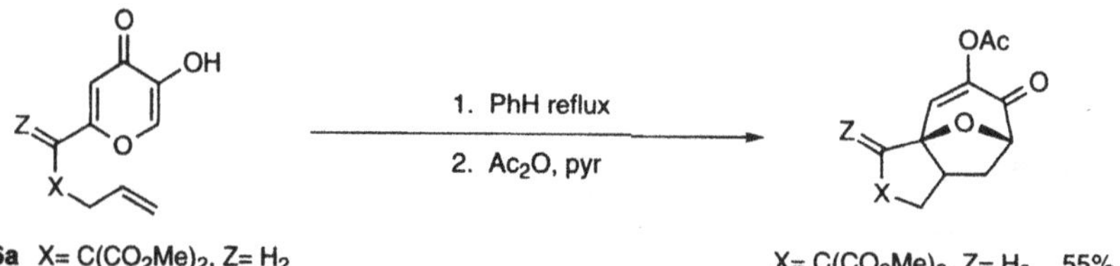

(33)

46a X= $C(CO_2Me)_2$, Z= H_2
46b X= MeN, Z= O

X= $C(CO_2Me)_2$, Z= H_2 55%
X= MeN, Z= O 70%

$$(34)$$

Wender's work in the intramolecular cycloadditions in the 4-pyrone series showed that the reaction is promoted by silyl group transfer, presumably via the complexation of the ketone to the electron-withdrawing silyl group [61]. Thus at elevated temperatures, substrate **48a** (R=TBDMS) underwent cycloaddition smoothly and stereoselectively while the analogous compound **48b** (R=Me) was inert, Eq. 35. The formation of the oxidopyrylium intermediate can also be promoted by the addition of methyl triflate [62]. Under these conditions, the intramolecular cycloaddition proceeded in one pot directly from the hydroxypyrone **49** at ambient temperatures to give **50**, Eq. 36. In contrast, only a trace amount of **50** was isolated from the thermolysis of **49**.

$$(35)$$

48a R= TBDMS
48b R= Me

$$(36)$$

Oxidopyrilium species may also be formed under photolysis conditions as demonstrated in Eq. 37. A high-energy diradical or a carbonyl ylide **52** produced upon photolysis of the epoxyketone **51** undergoes internal cycloaddition to give **53** [63].

$$(37)$$

Oxidopyrylium betaines from the 1,6-intramolecular addition of a carbene to the oxygen of a carbonyl group are presented along with other carbonyl ylides in the next section.

2.3
Cycloadditions of Cyclic Carbonyl Ylides

A review of the methods for the generation of cyclic carbonyl ylides from intramolecular carbene additions has recently appeared [64]. This intermediate was first exploited as the 4π component for cycloaddition reactions by Ibata [65]. *ortho*-Disubstituted carboalkoxy aryl diazoketones such as **54** were decomposed by copper complexes, generating six-membered ring carbonyl ylides. These transient intermediates underwent subsequent intermolecular cycloadditions in the presence of ethylenic and acetylenic reagents to give predominantly *exo* products containing the oxabicyclo[3.2.1] nucleus, Eq. 38.

$$\text{(38)}$$

Padwa subsequently described some intramolecular versions of this reaction. A structurally similar aryl ester, **55**, a tethered to a terminal olefin was subjected to rhodium-catalyzed diazo decomposition. Carbonyl ylide formation followed by intramolecular cycloaddition resulted in tricyclic product **56a**, Eq. 39 [66–69]. Cycloaddition also occurred with the amide analogue **55b**.

Carbonyl ylide formation in the aliphatic ketoester **57a** was complicated due to competing reaction processes; however, the diketone **57b** underwent cycloaddition at room temperature to give **58**, Eq. 40 [70].

$$\text{(39)}$$

55a X= O
55b X= NBn

56a X= O, 87% yield
56b X= NBn

$$\text{cat } Rh_2(OAc)_4, \text{ PhH, rt} \qquad 75\% \tag{40}$$

57a X= O
57b X= CH$_2$

58

Another example of an intramolecular cycloaddition was reported by Dauben, in which both isomers **59a** and **59b** of the substrate underwent tandem carbene-cyclization and intramolecular cycloaddition to give pentacyclic adducts **60a** and **60b**, Eq. 41. The stereochemistry of the newly constructed bonds was directed by the distal cyclopropane and was independent of the stereocenters at C_4, C_{10} or C_{12}.

The phorbol skeleton to yield **61** was also assembled in one step via rhodium catalyzed carbenoid formation, Eq. 42 [71].

$$\text{cat } Rh_2(OAc)_4, \text{PhMe, } 100°C, \text{ 1h} \qquad 86\% \text{ yield} \tag{41}$$

59a

59b

60a

60b

$$2 \text{ mol\% } Rh_2(OAc)_4, \text{ CH}_2Cl_2, \text{ rt, } 55\% \text{ yield} \tag{42}$$

61

In an analogous fashion, five-membered ring carbonyl ylides generated from diazodiones undergo cycloaddition with a variety of dipolarophiles, resulting in products containing the [2.2.1] oxabicyclic nucleus. These cycloadditions are generally highly regioselective and the *exo* isomer tends to predominate (Scheme 2). The reaction conditions are sufficiently mild that sensitive groups such as cyclopropyl substituents and acetals are tolerated [72–73].

A particularly interesting substrate bearing two diazoketones and two internal olefins was studied by Bien [40]. The course of the reaction involved the

Scheme 2

generation of a metallocarbene, cyclopropanation with one of the olefins, followed by carbonyl ylide formation and cycloaddition with the remaining olefin to yield pentacycle **62**, Eq. 43.

Padwa's experiments showed that five- and six-membered ring carbonyl ylide formation was facile while that of seven-membered carbonyl ylides was significantly more difficult, a reflection of the increased entropy of the system as a result of the lengthening of the chain [73]. Attempts to induce eight-membered ring carbonyl ylide formation were unsuccessful.

$$(43)$$

2.4
Fragmentation of Cyclic Oxonium Intermediates

In addition to the formation and reactions of carbonyl ylides discussed in the previous section, carbenoids also react intramolecularly with ethereal oxygen atoms to generate oxonium intermediates. When the ether is part of a ring as in substrates **63a–b**, the intramolecular addition of rhodium carbenoids produces bicyclic oxonium intermediates, which generated [5.2.1] oxabicycles **64a–b** upon rearrangement by a [2, 3]-sigmatropic pathway, Eq. 44 [74].

$$(44)$$

63a R= H
63b R= COOMe

64a R= H, 81% yield
64b R=CO$_2$Me, 67% yield

West has studied another system that generated oxonium intermediates which underwent fragmentation by a similar pathway, Eq. 45 [75].

For substrates lacking olefinic migrating groups, [1, 2]-shifts occur instead to give oxabicyclic products, Eq. 46.

$$(45)$$

R= H
R= Me

R= H   65% yield
R= Me   54% yield

$$(46)$$

15-19:1

2.5
Annulations of 1,3-Dinucleophiles with Dicarbonyl Compounds

Molander reported formation of [3.2.1] and [3.3.1] oxabicyclic compounds by applying Trost's trimethylenemethane dianion chemistry with dicarbonyl substrates, Eq. 47 [76]. The inherent symmetry of the dianion limits this methodology to those compounds with relatively simple and symmetrical structures.

In 1979, Chan reported the synthesis of oxabicyclo[3.2.1] and [3.3.1] compounds using the bis-silylated enol ethers of ketoesters **65**, which cyclized with dicarbonyl compounds and their tetrahydrofuran or tetrahydropyran derivatives in the presence of titanium tetrachloride, Eq. 48 [77].

Molander showed that for unsymmetrical dicarbonyl substrates the annulation is highly regioselective [76]. Complementary regioisomers of the cycloadduct can be obtained by variations in the reactivity of the dicarbonyl compound, Eqs. 49, 50.

By switching the Lewis acid to catalytic trimethylsilyl triflate or $TrSbCl_6$, the regioselectivity of the annulation with ketoaldehydes was reversed, showing that

$$(47)$$

100% yield, 10:1 diastereomers

$$(48)$$

65

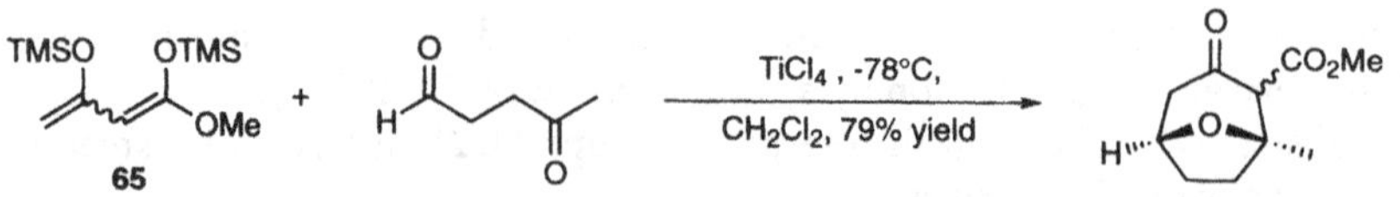

$$\text{(49)}$$

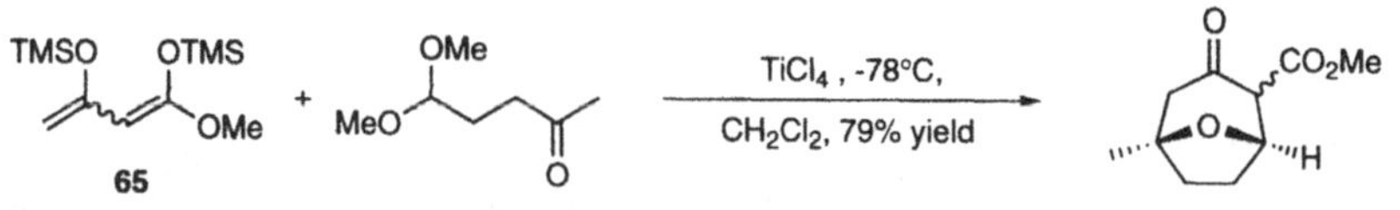

$$\text{(50)}$$

65 selectively reacted with the ketonic carbonyl group faster than the aldehydic carbonyl [78]. Even more impressive was the regioselectivity that was observed in the reaction of **65** with unsymmetrical diketo-substrates, Eq. 51.

A TMSOTf-initiated cyclization of the dicarbonyl substrate was invoked to explain the reactivity pattern [79]. Selective complexation of the less hindered carbonyl group activates it toward intramolecular nucleophilic attack by the more hindered carbonyl which leads to an oxocarbenium species. Subsequent attack by the enol ether results in addition to the more hindered carbonyl group. The formation of this cyclic intermediate also explains the high stereochemical induction by existing asymmetric centers in the substrates, as demonstrated by Eq. 52, where the stereochemistry at four centers is controlled. A similar reactivity pattern was observed for the bis-silyl enol ethers of β-diketones. The method is also efficient for the synthesis of oxabicyclo[3.3.1] substrates via 1.5-dicarbonyl compounds, as shown in Eq. 53. Rapid entry into more complex polycyclic annulation products is possible starting from cyclic dicarbonyl electrophiles [80].

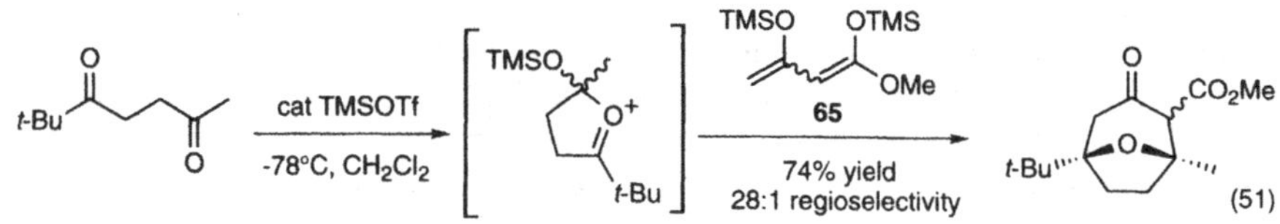

$$\text{(51)}$$

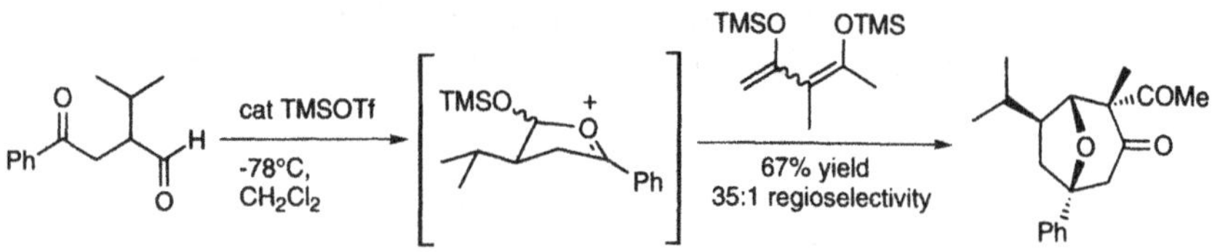

$$\text{(52)}$$

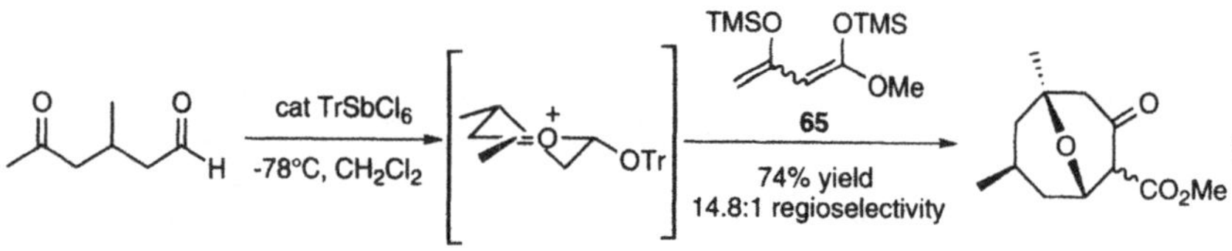

$$\text{(53)}$$

Interesting heteroatom-substituted derivatives such as **67** have also been synthesized via the reaction of bis enol ether **66** with thiol-containing dicarbonyl electrophiles, Eq. 54 [81]. Compound **68** bearing a bridgehead silyl substituent was produced from the reaction of **65** with a ketoacylsilane [82]. Subsequent decarboxylation and desilylation of **68** generates **69**, Eq. 55. The overall sequence represents a method to obtain the product of a formal inversion of the usual reactivity of **65** with ketoaldehydes. Extensive studies failed to reverse the observed regioselectivity.

2.6
Transannular Addition of Nucleophiles

Transannular additions by nucleophiles tend to occur with the larger carbocycles due to their increased flexibility, or in polycyclic compounds where reactive centers are forced into close proximity.

The generation of alkoxides in large-ring cycloalkanones results in the formation of relatively stable bicyclic hemiacetals. Thus, treatment of **70** with sodium borohydride results in the generation of **71a**, in equilibrium with its hydroxyoctanone tautomer, while treatment with phenyllithium gave the stable hemiacetal **71b** (Scheme 3). Similar reactions were observed with cycloheptane-1,4-dione and cyclooctane-1,5-dione; however, no transannular hemiacetal formation was observed for hydroxycyclohexanones [83].

Large cyclic hemiacetals are also formed from the intramolecular alkylation of ketal radicals of lactones, as demonstrated by Eq. 56 [84].

Under Lewis acidic conditions, cyclic acetals such as **72a–b** form oxonium ion intermediates which cyclize via an intramolecular Friedel-Crafts alkylation onto the tethered arene to form polycyclic benzylic ethers, Eq. 57 [85].

In electrophilic addition reactions, transannular participation by an oxirane to form an oxonium ion occurs in medium-ring cycloalkenes. Thomas first investigated this transannular addition in epoxide **73** [86]. Treatment with iodine gave iodinated bicyclic ethers **74** and **75**, Eq. 58.

Scheme 3

$$(56)$$

$$(57)$$

72a R= H
72b R= Me

R= H, 67% yield
R= Me, 96% yield

$$(58)$$

73   **74**   **75**

2 : 1

Subsequently, Martin showed that treatment of epoxyketone **76** with iodine gave predominantly **77**, as well as some **78**, Eq. 59 [87a].

Iodination of the 12-membered ring epoxide **79** gave an [8.2.1] oxabicyclic product, Eq. 60 [87b].

$$(59)$$

76   **77**   **78**

9 : 1

$$(60)$$

79

$$\text{(61)}$$

Lewis acid-induced ring-opening of **80** provided a synthesis of a [4.4.1] oxabicyclic system, Eq. 61 [88]. The product arises via a transannular nucleophilic opening by an internal epoxide.

2.7
Miscellaneous Reactions

Rigby synthesized a [5.3.1] oxa-bridged nucleus in the context of a perhydroazulene synthesis [89]. Upon treatment of **81** with BF3 etherate, a hetero-Diels-Alder reaction between the aldehyde and the diene occurred to give **82** in good yield, Eq. 62.

Gebel and Margaretha have reported a photochemical intramolecular [22] reaction between an olefin and a furanone which resulted in the construction of a [2.2.1] oxabicyclic nucleus as well as a cyclobutane in **83**, Eq. 63 [90].

Vogel has applied the Tiffeneau-Demjanov ring-expansion reaction to convert oxabicyclo[2.2.1] substrates into oxabicyclo[3.2.1] compounds [91]. The reduction of nitrile **4a** generates **84** which, under deamination conditions, yielded **85**, Eq. 64. Because **4a** can be obtained in enantiomerically pure form, this constitutes a enantioselective synthesis for oxabicyclo[3.2.1] substrate **85**.

$$\text{(62)}$$

$$\text{(63)}$$

$$\text{(64)}$$

3
Survey of Functionalization Reactions of Oxabicyclic Substrates

3.1
Stereoselectivity of Functionalizations

The inherent facial bias of many oxabicyclic compounds, as well as the stereo-
electronic influences that are exerted on remote groups as a consequence of
being constrained in a rigid bicyclic system, cause many functionalization reactions to occur with predictable and high levels of regioselectivity and stereoselectivity. The reactions on oxabicyclo[2.2.1] substrates have been the most
studied, and to a lesser extent the [3.2.1] oxabicyclic ring systems. These results
provide a basis for further investigations of analogous reactions on the less
studied large oxabicyclic ring systems.

3.1.1
exo/endo Selectivity

Functionalizations of 7-oxabicyclo[2.2.1]heptane systems have been particularly
well-documented. In the absence of substituents which have overwhelming
steric effects, the dominant selectivity is influenced by the oxygen bridge, which
steers the approach of reagents to the *exo* face of the substrate on the basis of
steric considerations as well as by chelation. This selectivity applies to a wide
variety of reactions on compounds with an olefinic bridge including osmylation
[92, 93], epoxidation [94], aziridination [95], hydrogenation [96], and hydroboration [97, 98]. The [3 + 2] cycloaddition with *C,N*-diphenylnitrone also generates
the *exo* adduct [99], as did the Pauson-Khand reaction with acetylene [100].
Addition reactions occur by initial formation of an *exo* intermediate, followed
by *endo* attack. Examples include chloroselenation [95c], bromoselenation [101]
and chlorosulfenylation [102]. Bromination follows the same course to give the
trans addition product, although leakage to the *cis* addition product by subsequent rearrangement has been observed [100, 103, 104].

Reactions of ethylenic bridged oxabicyclic ketones also undergo *exo* attack by
organolithium reagents [105], Grignard reagents [106], dichloroketene [107]
and hydridic reductions including LiAlH$_4$ [108], NaBH$_4$ [101, 109a] and aluminum isopropoxide [108].

In an analogous fashion, enolates of ethylenic bridged ketones such as **86** are
trapped predominantly from the *exo* face by methyl iodide, Eq. 65 [95, 109]. In
this case, the methylated ketone was subsequently reduced from the *endo* face
due to the hindrance of the *exo* substituent impeding the usual approach *syn* to
the bridgehead oxygen.

$$\text{(65)}$$

86 Ar=4-MeC$_6$H$_4$

Ager's studies on substrate **87** showed that electrophilic addition to the ester enolate occurs preferentially from the *exo* face, and the selectivity can be further improved by adjusting the reaction conditions (Scheme 4) [110].

The sole exception of preferential *endo* attack is seen in the reaction of cuprates with oxanorbornenyl ketones [106]. The unusual and unprecedented *endo* delivery of the nucleophile is proposed to proceed via a prior complexation of the bridgehead oxygen with one equivalent of the cuprate on the less hindered side, followed by addition of another equivalent of cuprate from the more hindered *endo* face of the carbonyl group. Table 2 shows the reactions of **88** with various cuprates to give the *exo* alcohols (entries 1–3). The remote olefin shows a positive effect in promoting *endo* nucleophilic attack, as shown by the reactions of **88** and **89** respectively (entry 1 vs. 4).

solvent	electrophile	exo: endo	yield
THF	MeI	2:1	76%
THF	PhSSPh	2:1	65%
Et$_2$O	PhSSPh	4:1	65%
hexane	PhSSPh	20:1	65%

Scheme 4

Table 2. Reaction of cuprates with oxanorbornyl ketones

Entry	Substrate	Cuprate	Endo attack :	Exo attack
1	**88**	Me$_2$CuLi		6 : 1
2	**88**	*n*-Bu$_2$CuLi	> 100 : 0	
3	**88**	Ph$_2$CuLi	6 : 1	
4	**89**	Me$_2$CuLi		2 : 1

Due to a strong preference for *exo* attack, indirect methods are required to install *endo* substituents. For example, addition reactions which proceed through bridged cationic intermediates cause other nucleophilic species to add from the *endo* face. Hence, *endo* halide substitution follows bromonium and selenonium ion formation. A striking example of this phenomenon is the internal nucleophilic addition of *endo* alkoxy groups as shown when the oxabicyclic acetal is treated with bromine, Eq. 66 [103]. Similarly, displacement of *exo* nucleofuges allows nucleophilic substitution from the *endo* face. This sequence has been demonstrated in the intramolecular nucleophilic ring opening reactions of *exo* epoxides and aziridines.

Another indirect route to *endo* substituents in the cycloadduct commences with an appropriately substituted furan derivative. After cycloaddition, the olefinic bridge is hydrogenated from the less hindered side, forcing the substituents into the *endo* position. This strategy was used to mimic a net dihydroxylation from the more hindered face, Eq. 67 [111].

$$(66)$$

$$(67)$$

Reactions on the three-carbon bridge of oxabicyclo[3.2.1] compounds have been reported but less systematically studied. Because the majority of these compounds are derived from oxyallyl cation cycloadditions, most experiments on the three-carbon bridge involve addition to the bicyclic ketone. The parent oxabicyclo[3.2.1] ketone **90** undergoes reduction with bulky hydride sources such as L-selectride to generate the *endo* alcohol, Eq. 68 [112]. Presumably, the selectivity is due to equatorial attack of the hydride at the ketone of the pyranone in a pseudo chair conformation. The *exo* alcohol is prepared from the *endo* alcohol by a Mitsunobu inversion-hydrolysis sequence [113].

Less bulky sources of hydride such as DIBAL-H or $NaBH_4$ lead to mixtures of *endo* and *exo* alcohols, Eqs. 69, 70 [100].

$$(68)$$

$$(69)$$

85:15

$$(70)$$

Alpha substituents which provide anchors for the pseudo-chair conformation increase the tendency for *exo* approach of reagents, Eqs. 71, 72 [114].

Substrates such as **91** which are locked in the pseudo-chair conformation, undergo exclusive *exo* addition of reagents, even with relatively small nucleophiles such as LiAlH$_4$, Eq. 73 [115].

$$(71)$$

$$(72)$$

$$(73)$$

Substituents which destabilize the chair conformation give stereoisomeric products, as demonstrated by the reduction of substrate **92**, which in contrast to the results above, exclusively generates the *exo* alcohol **93**, Eq. 74 [116]. Ketone **92** presumably exists in a boat conformation with pseudo-equatorial methyl groups rather than in a pseudo-chair conformation which would place the methyl substituents in pseudo-axial positions. Hydride attacks from the less-hindered *endo* face to provide the *exo* and pseudoaxial alcohol.

For substrates in which the carbonyl group is alpha to the bridgehead carbon in the three atom bridge, Grignard reagents [61, 117] and sodium borohydride [91] add *syn* with respect to the oxygen bridge. A detailed study by Sammes showed that an *exo* methyl group in the gamma position hindered the *exo* addition of hydride, leading to the production of *exo/endo* mixtures of alcohols, Eq. 75 [117b].

$$(74)$$

$$(75)$$

R= H   Yields: 92%   0%
R= Me   40%   30%

Alkylations of the enolate also occurred *syn* to the oxygen bridge, thus allowing the sequential functionalization of **94** to be achieved, Eq. 76 [59].

A variety of nucleophiles are added *syn* to the oxygen bridge in enone **95** including organocuprates and nucleophilic epoxidizing agents (Scheme 5) [118].

$$(76)$$

Scheme 5

3.1.2
Regioselectivity

Remote substituents have a dramatic effect on the regioselectivity of additions to oxabicyclo[2.2.1] systems as reported by Vogel. The results have been summarized [11, 119].

Ab initio calculations have revealed that the carbonyl group of **2** releases electron density to the olefin through homoconjugation [11d]. Therefore, of the various contributors of **96, 96b** predominates due to electron donation from the carbonyl group to the cation, which results in regiospecific addition to form **97** (Scheme 6). In contrast, the electron-withdrawing substituent nitrile in **98** makes **99a** the most important contributor by a repelling field effect. Regioisomeric addition products **100** are formed from **98**.

Scheme 6

These reactivity patterns have been observed in various electrophilic additions to oxabicyclo[2.2.1]heptenes (Scheme 7).

Oxabicyclo[3.2.1]octenone **85** initially forms the kinetic regioisomer **101**, but upon prolonged reaction time, the formation of thermodynamically favored regioisomer **102** is also observed, Eq. 77.

Scheme 7

$t^* = 0.5h, 13:1$
$t^* = 4h, 8:1$
$t^* = 8\ days, 1:1$

(77)

3.2
Enantioselective Desymmetrization Reactions

Asymmetric derivatization of *meso* oxabicyclic compounds generates enantiomerically enriched oxabicyclic compounds and provides a source of chiral oxabicyclic starting materials.

Table 3. Products and Methods of Enantioselective Desymmetrization by Esterification

Entry	Product	Substrate	Reaction	Yield[1]	ee[2]	Reference
1	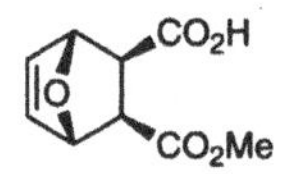 CO_2H, CO_2Me	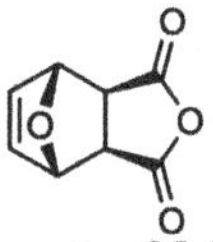 (anhydride)	cinchonidine, Et_2Zn THF, MeOH	- -	33%	[141b]
2	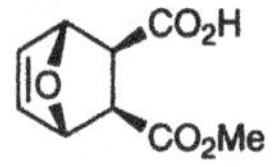 CO_2H, CO_2Me	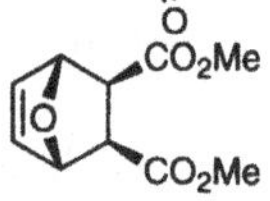 CO_2Me, CO_2Me	PLE[3], pH7 0.1M phosphate buffer	86% (61%)	75% (≥98%)	[141e]
3	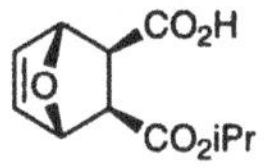 CO_2H, CO_2iPr	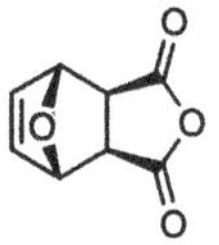 (anhydride)	TADDOLate **103**[4] THF	63%	98%	[141g]
4	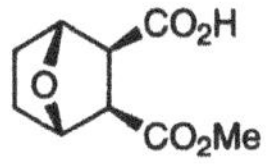 CO_2H, CO_2Me	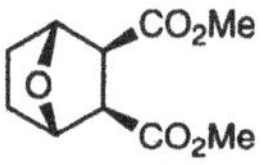 CO_2Me, CO_2Me	PLE[3], pH7 0.1M phosphate buffer	82%	≥98%	[141e]
5	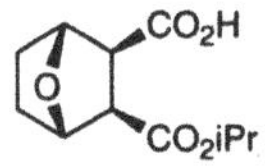 CO_2H, CO_2iPr	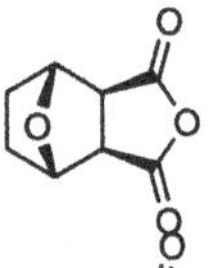 (anhydride)	TADDOLate **103**[4] THF	82%	96%	[141g]
6	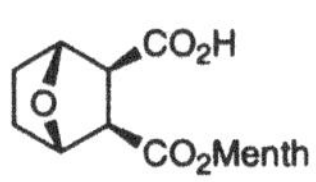 CO_2H, CO_2Menth	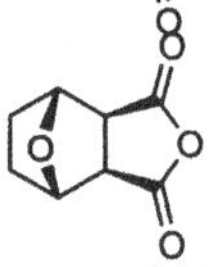 (anhydride)	*n*-BuLi, (*L*)-menthol THF, 78°C	26%	>98%	[141f]
7	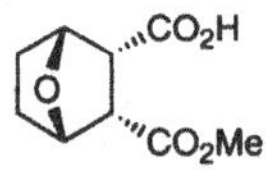 CO_2H, CO_2Me	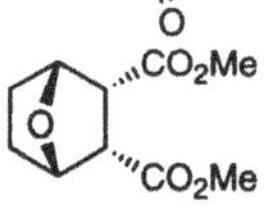 CO_2Me, CO_2Me	PLE[3], pH7 0.1M phosphate buffer	87% (65%)	64% (97%)	[141a]
8	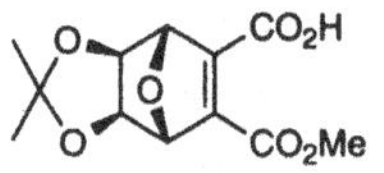 CO_2H, CO_2Me	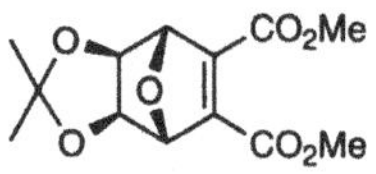 CO_2Me, CO_2Me	PLE[3], pH8 0.1M phosphate buffer	96%	77%	[141a]
9	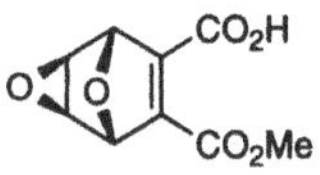 CO_2H, CO_2Me	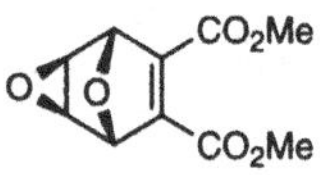 CO_2Me, CO_2Me	PLE[3], pH8 0.1M phosphate buffer	100%	77%	[141a]

[1] Yields after recrystallization in brackets.
[2] ee's after recrystallization in brackets.
[3] PLE = pig liver esterase.
[4] For structure of TADDOLate **103**, see text.

3.2.1
Desymmetrization of Oxabicyclo[2.2.1] Substrates

Table 3 lists the various methods that have been used to esterify or hydrolyze *meso* oxanorbornyl derivatives to provide enantiomerically enriched material. Very high enantiomeric excesses have been obtained using enzymatic techniques although this approach suffers from a lack of generality (entries 2,4). The desymmetrization using Seebach's TADDOLate **103**, Fig. 3, appears to be more tolerant of changes in remote functionality while maintaining high yields and enantiomeric excesses in the products (entries 3,5).

Table 4 shows reducing and oxidizing systems that have been used for desymmetrization.

Lautens and Ma made use of Brown's asymmetric hydroboration reaction to afford optically enriched alcohol **104** in 83% ee, Eq. 78 [120, 121].

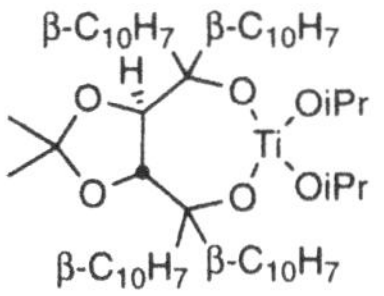

TADDOLate **103**

Fig. 3

Table 4. Products and Methods of Enantioselective Reductive or Oxidative Desymmetrization

Entry	Product	Substrate	Reaction	Yield	ee	Reference
1	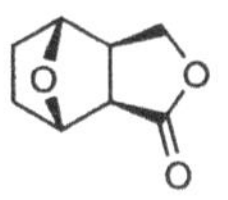	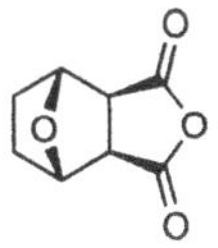	(+)-BINOL[1], LAH EtOH, THF -78°C	72%	83%	[141c]
2	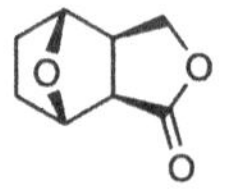	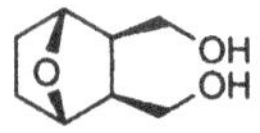	HLADH[2], pH 9 20% NAD[3], FMN[4] 10-24 days	83%	>98%	[141d]
3	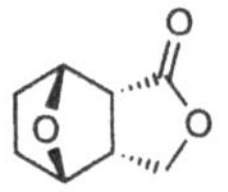		HLADH[2], pH 9 20% NAD[3], FMN[4] 10-24 days	37%	>98%	[141d]
4	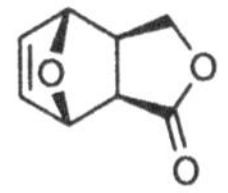	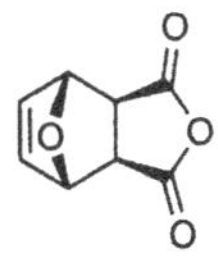	(+)-BINOL[1], LAH EtOH, THF -78°C	63%	99%	[141c]

[1] BINOL = 1,1'-bi-2-naphtol.
[2] HLADH = horse liver alcohol dehydrogenase.
[3] NAD = nicotinamide adenine dinucleotide.
[4] FMN = flavin mononucleotide.

$$(78)$$

3.2.2
Desymmetrization of Oxabicyclo[3.2.1] Substrates

Deprotonation of [3.2.1] oxabicyclic substrates by homochiral base **105** has been investigated by Simpkins [121]. Initial enantiomeric excesses of about 80 % can be achieved which are improved by successive recrystallizations. Table 5 summarizes these transformations.

Asymmetric hydroboration of substrates **106a** and **106b** yielded *exo* alcohols with high enantioselectivities, Eq. 79 [120].

Table 5. Enantioselective Desymmetrization by Deprotonation with Homochiral Base 105 [121]

105

Entry	Product	Substrate	Reaction	Yield	ee[1]
1			TMSCl THF, -94°C	88%	85%
2			1. TMSCl, THF, -95°C 2. PhIO, BF$_3$·Et$_2$O	59%	85% (≥98%)
3			TMSCl THF, -94°C	79%	88%

[1] ee after recrystallization in brackets.

$$(79)$$

106a R = H
106b R = Me

R = H   89% yield, 96% ee
R = Me 87% yield, 95% ee

4
Ring Opening Reactions of Oxabicyclic Substrates

This section describes the most commonly used methods used to cleave one or more bonds within the oxabicyclic framework. Aspects of this subject have been surveyed in other reviews describing the synthetic utility of [4+2] and [4+3] cycloaddition reactions [1, 25, 119, 122, 123]. The retro-Diels-Alder reactions of oxabicyclo[2.2.1] compounds under thermolytic conditions resulting in the extrusion of furan, acetylene or other stable species will not be covered, but the interested reader is directed to a recent review on this topic [122].

4.1
Cleavage of Carbon-Carbon Bonds in the Oxabicyclic Framework

Oxabicyclic substrates have been frequently used as precursors to highly substituted cyclic ethers, particularly tetrahydrofurans and tetrahydropyrans. This strategy relies on the selective cleavage of carbon-carbon bonds within the oxabicyclic nucleus.

4.1.1
Oxidation of the Carbonyl Functionality

Oxabicyclic substrates containing a carbonyl group can be readily cleaved by a Baeyer-Villiger oxidation-hydrolysis sequence. Vogel has performed an extensive study of the regioselectivity of this reaction in the context of oxabicyclo[2.2.1]heptanyl substrates, and has identified several useful trends [124a]. In the absence of special substituents and overwhelming steric effects, the oxidation product generally arises from migration of the bridgehead carbon. The enhanced migratory ability of the bridgehead carbon is attributed to the favorable through-bond interactions between the bridging oxygen and the carbonyl group [124b]. This pattern of reactivity permits the regioselective transformation of **107** to **108** which was used in the total synthesis of castanospermine and its deoxy-derivatives, Eq. 80 [103]. Moreover, when the alpha oxygen is protected as an ether, the directing effect is greater than that of an ester. Therefore, the regioselectivity of the Baeyer-Villiger oxidation can be influenced by the type of protecting groups that are used, as demonstrated in the reaction of **109** in Eq. 81. Subsequent elaboration of lactone **110** ultimately yielded D-lividosamine, an aminoglycoside antibiotic [125].

Table 6 shows additional reactivity trends of the Baeyer-Villiger oxidation. It has been observed that when the alpha substituent is a methoxy or a siloxy

mCPBA, CH2Cl2, NaHCO3, 96% yield

107 → **108** → (±)-Castanospermine   (80)

$$ \text{(81)} $$

109 → **110** (one regioisomer) → (+)-D-lividosamine hydrochloride

Table 6. Baeyer-Villiger oxidations of 7-oxabicyclo[2.2.1]heptanone derivatives with alpha oxygen substituents [a]

Entry	Substrate	Product ratios		
1	X, Y, Z= H	>97:3		
2	X= OBz, Y, Z= H	7.8:1		
3	X= OMe, Y, Z= H	1:2.8		
4	X, Z= H, Y= OMe	<3:97		
5	X, Z= H, Y= OTBDMS	<3:97		
6	X= OMe, Y= H, Z= OMe	<3:97		
7	X= OBz, Y= H, Z= OBn	>95:5		

[a] Reaction conditions: mCPBA in $CDCl_3$ with $NaHCO_3$

group, the regioselectivity of the oxidation reverses (entries 3, 5). This selectivity is more pronounced for substrates where the alpha alkoxy substituent is *endo*, (entry 4) and the effect is further reinforced by the presence of an additional *endo* ether at the pseudo-*para* position (entry 6). In fact, the regioselectivity observed for an alpha ester is also enhanced by the presence of an *endo* alkoxy group (entry 7). While the origins of these effects are not well-understood, these studies provide a basis for predicting and using this reaction to selectively manipulate [2.2.1] oxabicyclic compounds in synthesis.

The Baeyer-Villiger ring cleavage of both [2.2.1] and [3.2.1] oxabicyclic compounds has been used as a key step in the synthesis of many natural products, including showdowmycin [126], nonactic acid [127], lilac alcohol [128], the C_{21} to C_{27} subunit of rifamycin [198], and various C-nucleosides [129]. The "naked sugar" substrates synthesized by Vogel have been used in the synthesis of many natural and unnatural sugars, as well as their derivatives, including D- and L-allose, D- and L-talose [95a, b], allonojirimycin [130], L-daunosamine [95c], and various disaccharides [131].

Vogel also used (−)-5, prepared from 2,4-dimethylfuran, to show that a sequence involving stereoselective functionalization, fragmentation via Baeyer-Villiger oxidation and exhaustive reduction constitutes a quick assembly of optically pure polypropionate arrays with four contiguous stereocenters, Eq. 82 [14, 132].

A variation of this strategy employed the Beckmann rearrangement to insert nitrogen, eventually leading to a synthesis of a muscarine analogue, Eq. 83 [133].

Ring scission which is complementary to that using the Baeyer-Villiger oxidation-hydrolysis sequence has also been developed. The "naked sugar" derivative **111** was converted to its silyl enol ether and then ozonolyzed and reduced to give **112** which was transformed to the C-nucleoside, cordecepin C, Eq. 84 [109a].

Ketone **113** was similarly cleaved via its corresponding silyl enol ether, eventually leading to the synthesis of tiazofurin, a potent antiviral and antitumor agent, Eq. 85 [134].

Oxabicyclic ketones have also been further derivatized to the alpha-oxidation products which are in turn cleaved, offering still another option for carbon-carbon bond scission. For example, hydroxyketone **114**, available in >98% ee from the parent ketone, was cleaved by lead tetraacetate to afford an excellent yield of the hydroxyester **115**, a key intermediate in Noyori's synthesis of showdomycin, Eq. 86 [121, 129]. In this case, the ozonolysis of the silyl enol ether of the parent ketone led to complex mixtures, demonstrating the complementarity of these approaches.

$$\text{114} \quad \xrightarrow[\substack{\text{2. NaCNBH}_3 \\ \text{93\% yield}}]{\text{1. Pb(OAc)}_4\text{, MeOH}} \quad \text{115} \qquad (86)$$

4.1.2
Oxidative Cleavage of Vicinal Diols in the Carbon Framework

Periodate promoted cleavage of vicinal diols has also been used to prepare monocyclic products. Oxabicyclo[4.2.1]nonadiene **116** derived from diiodoketone **77** was subjected to sodium periodate and sodium borohydride reduction to generate **117**, Eq. 87. Subsequent elaborations resulted in the stereocontrolled synthesis of oxepine **118**, a subunit designed for the assembly of polyether toxins such as ciguatoxin [135].

$$\text{77} \quad \longrightarrow \quad \text{116} \quad \xrightarrow[\substack{\text{2. NaBH}_4 \\ \text{68\% yield}}]{\text{1. NaIO}_4} \quad \text{117} \quad \longrightarrow \quad \text{118} \qquad (87)$$

Periodate cleavage of dihydroxy oxabicyclic substrate **119** generated an unsymmetrical subunit useful for polyether assembly, Eq. 88 [88].

Periodate cleavage of an oxabicyclic diol was also a key step in the synthesis of citreoviral from the Diels-Alder adduct of 2,4-dimethylfuran and vinylene carbonate [136].

$$\text{119} \quad \xrightarrow[\substack{\text{2. ethylene glycol, PhH, CSA, 12 h } \Delta \\ \text{93\% yield}}]{\text{1. NaIO}_4\text{, MeOH, H}_2\text{O, rt, 4h}} \qquad (88)$$

4.1.3
Oxidative Cleavage of the Carbon Framework

Carbon-carbon double bonds in oxabicyclic systems are cleaved by ozonolysis. Moreover, tri-substituted olefins generate cyclic ethers bearing side chains with differentiated ends upon ozonolytic cleavage, thus allowing subsequent selective elaboration of each appendage. Naked sugars were used extensively by Vogel as furanosides and C-nucleoside derivatives [11a].

In addition to the studies in the [2.2.1] oxabicyclic series, Vogel also subjected the [3.2.1] oxabicyclic vinyl chloride **120** to ozonolysis to produce a dialkylated tetrahydropyran **121** with differentially oxidized substituents at C_2 and C_6, Eq. 89 [91]. This sequence of reactions was utilized in the synthesis of β-C-hexopyranosides such as **122**.

Bicyclic ether **124**, obtained from the intramolecular cycloaddition of **123**, was subjected to ozonolysis with a reductive work-up, Eq. 90. Silyl protection gave alcohol **125** and reduction transformed the thioether linkage into the vicinal *cis* dimethyl groups found in (±)-nemorensic acid [137].

Previous synthetic studies that have employed ozonolysis as a means for cleaving oxabicyclic substrates include Meinwald's studies toward pederin [138], Just's synthesis of showdowmycin [139], Masamune's synthesis of avenaciolide [140], and Ohno's asymmetric syntheses of (+)-showdowmycin, (−)-cordycepin C, and (−)-6-azapseudouridine [141a].

A key step in the recent synthesis of (+)-lauthisan by Cha was the ozonolytic cleavage of the olefinic bond of the tricyclic substrate **127** to afford the cyclic ether **128**, Eq. 91 [115]. A series of transformations including an enzymatic desymmetrization completed the total synthesis.

Unsaturated oxabicyclic substrates can also be cleaved through their vicinal diol derivatives, as exemplified by the reaction of substrate **129**, Eq. 92 [87].

(91)

(92)

4.1.4
Retro-Dieckmann/Retro-Aldol Reactions

Oxabicyclic substrates containing a 1,3-dicarbonyl functionality have been ring-opened via a retro-Dieckmann reaction, whereas compounds bearing a β-hydroxycarbonyl motif undergo a retro-aldol ring cleavage. The driving force for these reactions to occur in oxabicyclic systems is the relief of ring strain present in the bicyclic framework.

In Kozikowski's synthesis of showdomycin, treatment of the oxabicyclic **130** with bicarbonate induced a retro-Dieckmann reaction to reveal the highly substituted tetrahydrofuran intermediate **131**, Eq. 93 [142].

Similarly, treatment of substrate **132** with sodium methoxide led to a retro-Dieckmann reaction to yield the interesting bicyclic ether **133** as a single isomer, Eq. 94 [90].

A 2-siloxyfuran bearing a chiral auxiliary underwent a facially selective [4+2] cycloaddition to give **134**. Desilylation using fluoride gave furanone **135** via a retro-aldol reaction (Scheme 8). It is interesting to note that treatment of the same substrate with PPTS led to oxygen bridge cleavage to give hydroxycyclohexenone **136** [143].

(93)

(94)

Scheme 8

Katagiri's group has developed a reductive retro-aldol reaction to cleave [2.2.1] oxabicyclic substrates bearing *gem*-diesters [111, 144]. The acetate **137** underwent a retro-aldol reaction to afford a quantitative yield of a C-nucleoside precursor **138**, Eq. 95.

$$K_2CO_3, NaBH_4, MeOH \quad \text{quantitative yield} \tag{95}$$

4.1.5
Photochemically-Induced Cleavage

Photolysis of the hypoiodite of hemiacetal **139** results in carbon-carbon bond cleavage to produce an iodolactone **140**, Eq. 96 [83]. The iodoalkyl side chain was subsequently homologated to afford the sex pheromone of the rove beetle.

Padwa has introduced a rearrangement of oxabicyclic substrates that efficiently assembles oxa-polyquinane derivatives [145]. Oxabicyclo[3.2.1]alkenes **141** bearing a carbonyl group alpha to the bridgehead position can undergo a facile photoinduced 1,3-sigmatropic rearrangement. Thus the photolysis of **141** affords the linear oxatriquinane **142**, Eq. 97, while **143** generates the angular oxatriquinane **144** , Eq. 98. Both substrates **141** and **143** were obtained via the rhodium-catalyzed tandem-cyclization cycloaddition developed by Padwa.

$$\tag{96}$$

(97)

(98)

4.1.6
Electrochemical Cleavage

Akiyama's group developed an anodic oxidative decarboxylation of oxabicyclo[2.2.1] substrates that subsequently undergo skeletal rearrangement to yield 1,2,3-trisubstituted cyclopentanols [146, 147]. An example of this reaction which generates the carbocyclic framework of hydrindanes is shown in Eq. 99.

(99)

4.1.7
Acid-Induced Skeletal Rearrangements

The oxygen bridge in oxabicyclic compounds is an electron pair donor that can stabilize α-carbocations. This characteristic renders oxabicyclic substrates more susceptible to carbocationic skeletal rearrangements resulting in the cleavage of the carbon framework. One such reaction was exploited by Sammes for the synthesis of (±)-cryptofauronol, in which treatment of 145 with Lewis acid induces rearrangement to a decalin ring system, Eq. 100 [57].

(100)

A similar rearrangement to a [4.4.0] carbocyclic skeleton was observed by Harmata upon treatment of **146** with bromine. The proposed mechanism involves formation of a bromonium ion which rearranges and loses a proton to form an enol ether, which reacts with a second mole of bromine to give, after hydrolysis, an excellent yield of the rearranged product (Scheme 9) [148].

The epoxidized oxanorbornane derivatives **147** and **148** also rearranged under acidic conditions [94]. Remote substituents direct the cleavage of the carbon-carbon bonds (Scheme 10).

Scheme 9

Scheme 10

4.1.8
Miscellaneous Cleavage Reactions

Sodium naphthalide induced fragmentation and ring opening in oxatricyclic substrate **149**, Eq. 101. The allylic sulfate formed underwent elimination to produce an oxabicyclo[6.2.1] system containing a *trans* olefin. Simple reduction and elimination of sulfate led to the minor product [149].

$$(101)$$

4.2
Cleavage of Carbon-Oxygen Bonds in the Oxabicyclic Framework

The cleavage of the ether bonds in the oxabicyclic framework has been developed into a useful strategy to generate highly-substituted cyclohexyl derivatives from oxabicyclo [2.2.1] systems. This is an attractive approach because the facial bias inherent to the oxabicyclic substrate can be exploited to control the stereochemistry before ring opening is induced.

Functionalized medium-sized carbocyclic rings (especially seven- and eight-membered rings) can be accessed by this route through the use of the appropriate [3.2.1], [3.3.1], or [4.2.1] oxabicyclic substrates. This approach avoids otherwise entropically disfavored cyclization approaches to these rings, and therefore has proven particularly valuable for the syntheses of certain families of natural products.

In addition, further cleavage of the monocyclic products resulting from bicyclic ring cleavage affords efficient routes to polysubstituted acyclic chains, which are otherwise obtained from sequential coupling of smaller building blocks. In view of the synthetic potential underlying the selective cleavage of the carbon-oxygen bonds in the oxabicyclic framework, the development of this methodology is an attractive approach. Mounting interest in the utility of this strategy has been evident in the increasing number of recent publications.

4.2.1
Oxygen Bridge Activation by an Electron-Donating Group at the Bridgehead Carbon

The cleavage of the ether bond in an oxabicyclic compound bearing an oxygen substituent at the bridgehead (i.e. a bicyclic ketal or hemiketal) can be readily accomplished by solvolysis under acidic or basic conditions.

The oxabicyclic compounds required for this sequence can be assembled using 2-oxygenated furans. Gravel and Brisse synthesized the oxabicyclic substrate 150 by the Diels-Alder reaction between 2-acetoxyfuran and chloromethyl maleic anhydride [150]. Ring-opening was induced by treatment with methanolic hydrogen chloride to give hydroxycyclohexanone 151, Eq. 102.

$$(102)$$

Chiral cycloadduct **134** assembled from a tethered 2-siloxyfuran was treated with PPTS to reveal the hydroxycyclohexenone **136** (Scheme 8) [143]. Other natural products that have been synthesized employing this strategy include triptonide and triptolide [151].

Recently, the first total synthesis of taxodone was accomplished via this strategy [152]. Cycloadduct **153**, readily available from the Diels-Alder reaction of siloxyfuran **152** and methyl acrylate, was treated with acid to induce ring opening and dehydration to afford phenol **154**, Eq. 103.

Alternatively, a bicyclic hemiketal can be unmasked just prior to hydrolysis. This strategy was cleverly applied by White to his synthesis of the Prelog-Djerassi lactone [114]. Instead of carrying a potentially labile acetylated hemiketal, White began the synthesis using 2-acetylfuran, from which oxabicyclic substrate **155** was obtained. The hemiketal functionality was created by a Baeyer-Villiger oxidation-basic hydrolysis sequence which resulted in ring opening to give the hydroxyheptanone **156**, Eq. 104.

4.2.2
Generation of a Carbanion α to the Carbon-Oxygen Bond

Several strategies for ring opening are based on the elimination of alkoxide by the generation of a carbanion alpha to the bridgehead position.

This carbanion can be readily generated in an oxabicyclic compound **157** bearing an electron-withdrawing group on the carbon alpha to the bridgehead, Fig. 4. Alternatively, an electron-withdrawing functional group built into the oxabicyclic structure as in **158** would also facilitate the formation of the required carbanion. Treatment with base results in ring opening via an elimination.

Base-induced ring openings of this kind have been used extensively for the preparation of many natural products. Several of the syntheses of shikimic acid

Fig. 4

derivatives have utilized this approach [17b, 153–158], as did the earlier work on 11-ketotestosterone [152], and gibberellic acid [159]. For example, the enantiomerically pure oxabicyclo[2.2.1] substrate **159** was treated with LHMDS to give the ring opened cyclohexenol **160**, which yields (+)-5-epi-methyl shikimate **161** upon deprotection, Eq. 105 [96]. Subsequent reactions have transformed **161** into several optically active pseudo-sugars [160].

LHMDS, THF, -78°C, 0.5h, 58%

AcOH, H$_2$O, 55°C, 3h, 96%

(105)

159 → **160** → **161**
(+)-5-epi-methyl shikimate

The less-hindered acidic proton in **162** was deprotonated selectively to afford tertiary alcohol **163**, Eq. 106 [72].

The base-induced ring opening of **164** gave **165** which was used in an efficient, six-step synthesis of illudin M, Eq. 107 [161].

LDA, -78°C, 78% yield

(106)

162 → **163**

KOH, MeOH, 42% yield

(107)

164 → **165** → illudin M

The extremely labile bacterial oxidation product of phthalic acid, 4,5-*cis*-dihydrodiol **167** was synthesized via the base-induced ring opening of oxabicyclo[2.2.1] substrate **166**. Selective deprotonation of the less-hindered *exo* proton was possible, Eq. 108 [162].

The differentially protected dialdehyde **168** also underwent efficient ring opening under basic conditions, Eq. 109.

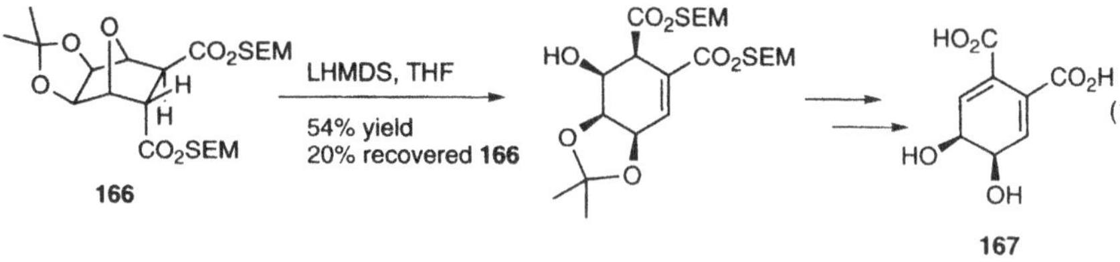

The analogous ring opening of the sulfonylated derivatives of [2.2.1] oxabicyclic compounds also proceeded to give cyclohexenyl sulfones as products [163]. Arjona exploited this reaction in the synthesis of pseudosugars [164]. When oxabicyclic sulfone **169** was treated with *n*-BuLi, selective ring opening of the bridging C–O bond to give **170** was observed rather than elimination of the *β*-benzyloxide, Eq. 110. After directing the ring opening, the sulfone was conveniently removed and **170** was dihydroxylated to give carba-*α*-DL-glucopyranose.

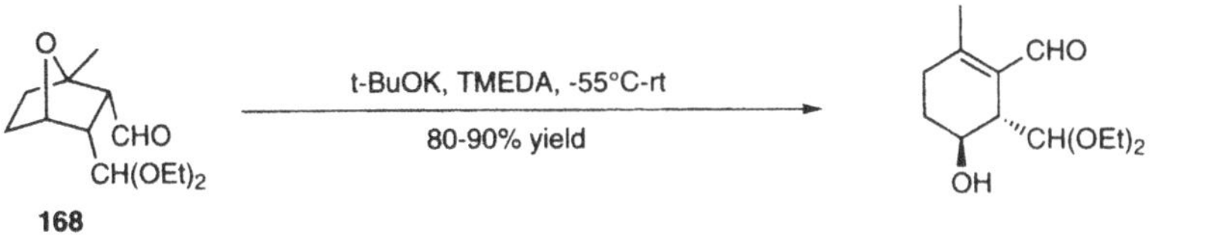

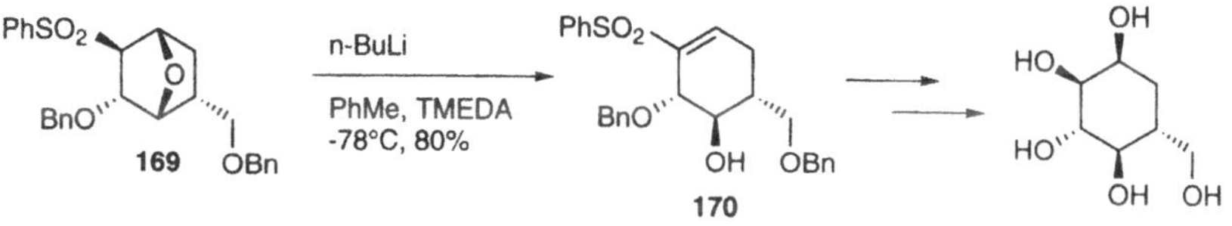

The stereocenters set in the Diels-Alder sultone cycloadduct **171** were unraveled by base-induced ring opening to afford hydroxysultone **172** in excellent yield, Eq. 111 [165]. Subsequent manipulations led to a synthesis of ivangulin.

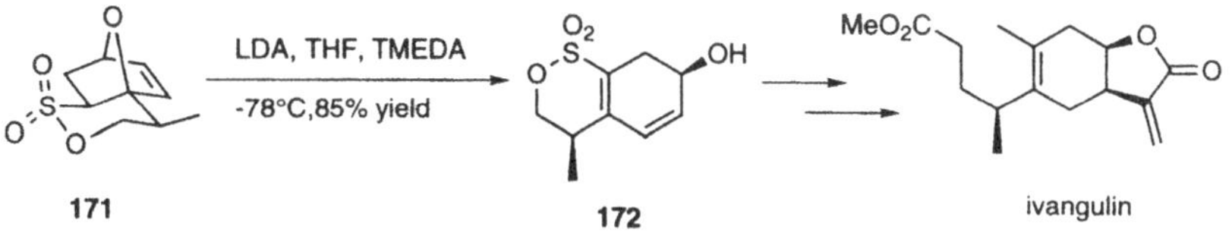

Deprotonation of an allylic proton in both isomers of bicyclic ether **173a** and **173b** using Schlosser's base led to dienol **174**, Eq. 112 [166].

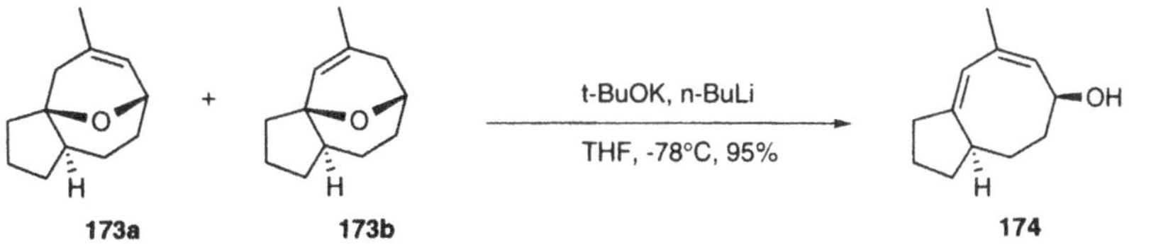

The allylic proton of the *exo* methylene derivative 175 was abstracted when treated with an organolithium reagent and subsequent elimination afforded dienediol 176, Eq. 113. The analogous ring opening reaction occurred for *exo* methylene [2.2.1] oxabicyclic substrates as well [120a].

$$175 \quad \xrightarrow[\text{84\% yield}]{\text{10 eqvts n-BuLi, THF, -61°C, 1 h}} \quad 176 \tag{113}$$

Weak bases can also induce ring opening with the aid of an oxaphilic reagent. Thus, the oxygen bridge in oxabicyclo[2.2.1]heptanone 177 was cleaved in the presence of triethylamine and TMSOTf to generate enone 178, which was an intermediate in the first total synthesis of (–)-conduritol C, Eq. 114 [93]. TBDMSOTf/triethylamine is also an effective combination for this transformation and has been used in the synthesis of *myo*-inositol derivatives, as well as (–)-conduritol B from 179, Eq. 115 [167, 168]. (+)-Conduritol F has also been prepared by this route which served to confirm its structure and demonstrate it was identical to natural (+)-leucanthemitol [168], Fig. 5.

$$177 \quad \xrightarrow{\text{TMSOTf, Et}_3\text{N}} \quad 178 \quad \longrightarrow \quad \text{(—)-conduritol C} \tag{114}$$

$$179 \quad \xrightarrow[\text{PhH, rt}]{\text{TBDMSOTf, Et}_3\text{N}} \quad \longrightarrow \quad \text{protected myo-inositol} \tag{115}$$

(-)-conduritol B   (+)-conduritol F

Fig. 5

4.2.3
Generation of a Carbanion γ to the Carbon-Oxygen Bond

The generation of a carbanion gamma to the oxygen bridgehead could also lead to elimination and ring opening. Arjona and co-workers explored this strategy in the cycloadducts of 2,4-dimethylfuran, the "naked sugars of the second generation" [169]. Treatment of **180** with LDA generated cyclohexadienol **181** in good yield, Eq. 116.

$$\text{(116)}$$

180   **181**

4.2.4
Heterolytic Cleavage Induced By Acids

Treatment of oxabicyclic compounds with strong acids can lead to the heterolytic cleavage of the ether bridge. The carbocationic intermediate can subsequently lose a proton to form an olefin, or react with a nucleophile. Since the reaction conditions are typically rather harsh, problems of chemoselectivity in the presence of sensitive functional groups as well as regioselectivity of the cleavage step are important issues. Rearrangement of the carbocationic intermediates can also potentially pose problems. Reagents which are useful for the cleavage of "typical" ethers have been used for the ring opening of oxabicyclic compounds [170] but the specific structure of the substrate frequently determines the outcome of the reaction, and not all reagents can be uniformly applied to all substrates [171].

It is particularly difficult to carry out a ring opening in compounds containing the oxabicyclo[2.2.1] nucleus without concomitant aromatization, because the strong acidic conditions can also lead to dehydration. On the other hand, inducing aromatization under controlled conditions permits the synthesis of highly-substituted aryl compounds as an alternative synthetic strategy to "traditional electrophilic aromatic substitution". The methods to aromatize oxabicyclo[2.2.1] heptanyl derivatives by the use of acids and low valent metals have been reviewed [172].

For the heterolytic cleavage of the bridging ether in oxabicyclo[3.2.1] substrates, it is essential to find reaction conditions to induce a regioselective opening, as well as complementary conditions selective for troponization, in light of the biological activity of many troponoids.

4.2.4.1
Protic Acids

A recent series of detailed mechanistic studies on the acid-catalyzed hydrolysis of 7-oxabicyclo[2.2.1]heptane derivatives **182**, **183** and **184** have confirmed that the reaction is initiated by protonation of the oxygen bridge, followed by a rate-limiting carbon-oxygen bond rupture to give a carbocationic intermediate. There are varying degrees of solvent assistance in the rate limiting step depending on the substrate [173], Fig. 6.

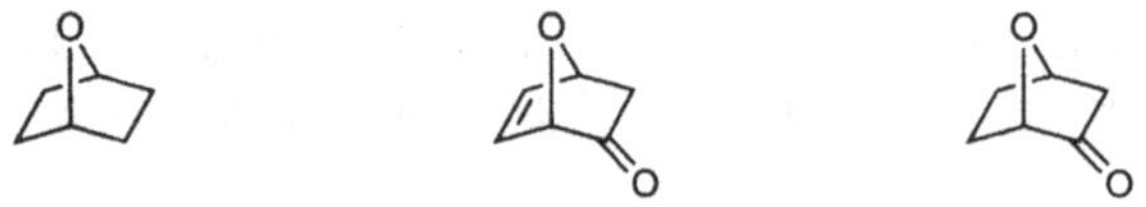

182   **183**   **184**

Fig. 6

Yates observed an intramolecular ring opening of **185** when it was subjected to treatment with formic acid, Eq. 117 [174].

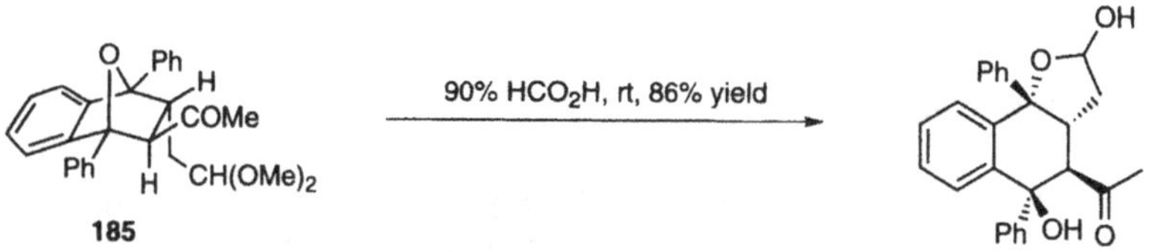

$$90\% \text{ HCO}_2\text{H, rt, 86\% yield} \tag{117}$$

185

Ogawa reported that although the acetolysis of **186** resulted in a mixture of pentaacetates from non-selective bridge opening, ring cleavage using hydrobromic acid generated a single dibromide **187**, which is the product of substitution at the less hindered bridgehead carbon by bromide, Eq. 118 [175]. The dibromide was eventually converted to the penta-N,O-acetyl derivative of (+)-validamine. Similarly, acidic treatment of **188** resulted in the exclusive formation of **189**. Dibromide **189** was also an intermediate in the syntheses of analogs of valienamine, Eq. 119 [176].

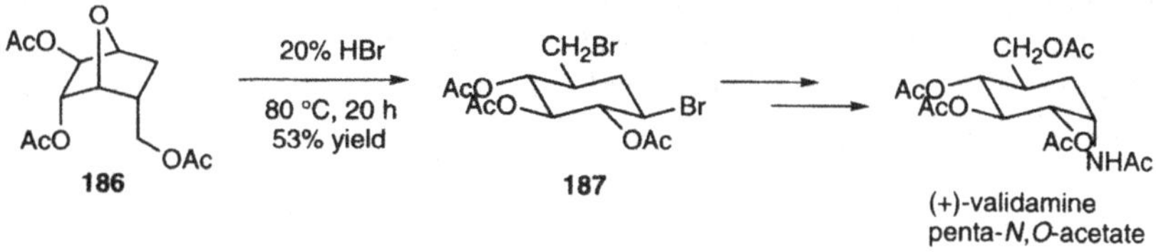

$$\begin{array}{ccc} 20\% \text{ HBr} & & \\ 80\,^\circ\text{C, 20 h} & & \\ 53\% \text{ yield} & & \end{array} \tag{118}$$

186   **187**

(+)-validamine
penta-N,O-acetate

$$\begin{array}{ccc} 18\% \text{ HBr, HOAc} & & \\ 80\,^\circ\text{C, 20 h} & & \\ 73\% \text{ yield} & & \end{array} \tag{119}$$

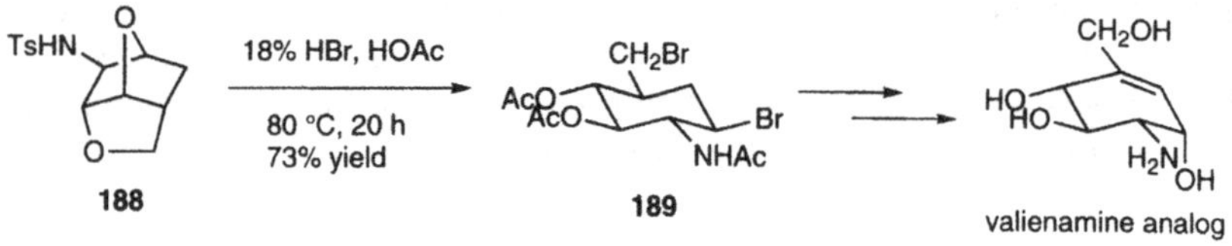

188   **189**

valienamine analog

The precursor of the hexasubstituted benzene in jatropholone A and B was a Diels-Alder adduct formed when furan **190** and enone **191** were reacted under high pressure. Subsequent aromatization was initiated by treatment with dilute hydrochloric acid, Eq. 120 [177]. This strategy was also used to install the aromatic ring in the syntheses of mansonone E [178].

Methoxy-substituted dihydronapthalene oxides undergo regiospecific oxygen bridge cleavage under acidic conditions, the selectivity of which is directed

$$\text{(120)}$$

jatropholone A, β-Me
jatropholone B, α-Me

by the formation of the carbocation that is stabilized by the methoxy group. Therefore, treatment of **192** under acidic conditions generated naphthol **193** in high yield, Eq. 121 [179].

$$\text{(121)}$$

When naphthalene oxides of general structure **194** are subjected to acid-induced ring opening, the generation of 2-substituted naphthols **196** were overwhelmingly favored over the 3-substituted naphthols (Scheme 11) [180]. The explanation put forward was that allylic cation **195** was significantly more stable than **197**.

196a:198a   98:2
196b:198b   >99:1
196c:198c   >99:1

194a R= Bn   Yields: 100%
194b R= Me   85%
194c R= Br   75%

Scheme 11

Noyori has transformed 8-oxabicyclooctanones into various naturally-occurring troponoids by acid-induced cleavage of the oxygen bridge, followed by dehydration and oxidation [181]. Equation 122 shows the synthesis of nezukone by this methodology. Although ether cleavage could be induced by boron trifluoride, fluorosulfuric acid was found to be the reagent of choice. Other troponoids such as hinokitiol and α-thujaplicin were synthesized by a similar strategy.

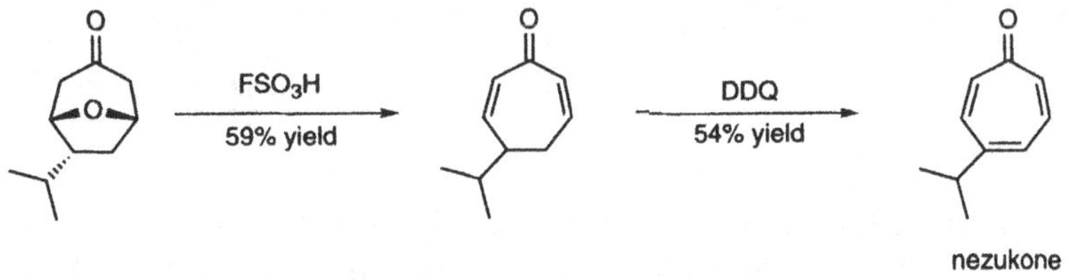

$$\text{(122)}$$

4.2.4.2
Boron-Based Lewis Acids

Lewis acids based on boron are effective reagents for the cleavage of "simple" ethers and have also been used to induce ring opening in many oxabicyclic substrates.

Kato treated **199** with boron trifluoride and acetic anhydride to achieve ring opening, Eq. 123. Furthermore, treatment of **200** with acetic toluene-p-sulfonic anhydride resulted in a regiospecific elimination to give **201** in quantitative yield, Eq. 124. Compound **201** was used as a precursor for ring A in synthetic studies toward fujenoic acid [182].

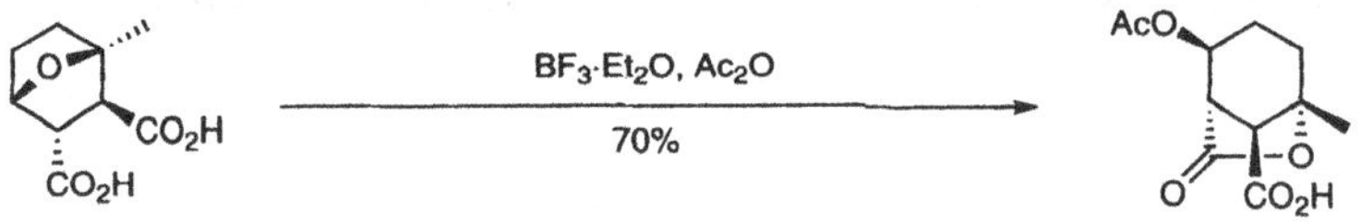

$$\text{(123)}$$

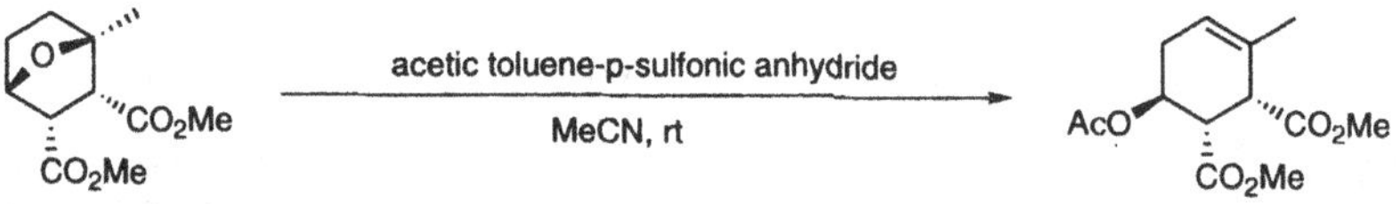

$$\text{(124)}$$

Whalley found that treatment of **202** with BCl$_3$ led to O-demethylation as well as regioselective heterolytic opening of the oxabicyclic nucleus [183]. The carbocation that is generated is trapped either by chloride to give **203** or by intramolecular cyclization by the phenol oxygen to give xanthone **204**, Eq. 125.

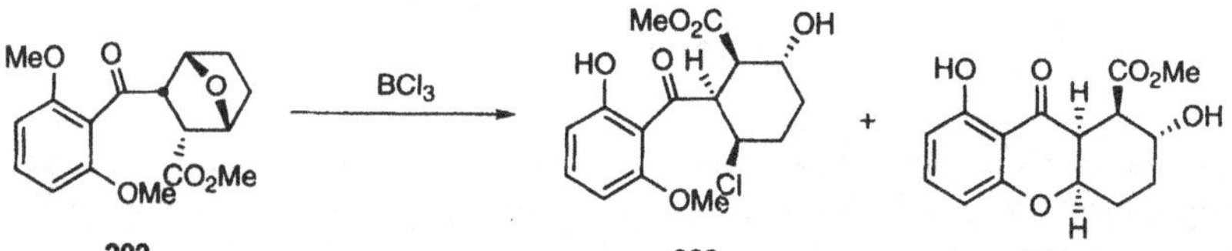

$$\text{(125)}$$

Recently, Koreeda observed a highly regioselective ring opening in connection with the synthesis of **205**, the carcinogenic *anti*-diol epoxide of benzo[a]pyrene [184]. The synthesis of the original oxabicyclic substrate was based on the [4+2] cycloaddition of the requisite aryne with 3,4-dibenzyloxyfuran. Following a series of high-yielding manipulations to obtain the cyclic carbonate **206**, treatment with BBr3 gave **207a**, Eq. 126. The regioselectivity observed agreed with theoretical calculations which indicate the stability of the bay region benzylic carbocation is higher than its non-bay region counterpart. However, the subsequent rapid epimerization at the brominated carbon of **207a** could not be prevented. In contrast, treatment with BCl₃ led to the analogous chloride **207b** which was sufficiently stable that it could be treated with aqueous base to complete the synthesis of the target. This methodology was also used in the synthesis of the *anti*-diol epoxides of 7,12-dimethylbenz[a]anthracene.

$$\text{(126)}$$

206 **207a** X= Br **207b** X= Cl **205**

Nicholas found that treatment of **208** with Me₂BBr results in elimination and ring opening to give a 9:1 mixture of dienes **209** and **210**, Eq. 127 [186].

$$\text{(127)}$$

208 **209** + **210**

9 : 1

Moreover, thioketalization in the presence of BF₃ etherate induced **211** to undergo addition-ring opening to afford olefin **212** regioselectively and in high yield, Eq. 128. This product was subsequently converted into damsin [186].

$$\text{(128)}$$

211 **212** damsin

Rigby also observed ring opening under similar conditions with [5.3.1] oxabicyclic substrate **82** [89]. However, the thiol nucleophile underwent both S_N2 and S_N2' addition to give a 2:1 mixture of **213** and **214**, Eq. 129.

$$\text{82} \quad \xrightarrow[\text{62\% yield}]{\text{PhSH, BF}_3\cdot\text{Et}_2\text{O}} \quad \text{213} \; + \; \text{214} \qquad (129)$$

$$2 : 1$$

4.2.4.3
Silyl Lewis Acids

Mann observed a regioselective ring opening of oxabicyclic substrate **215** using TMSI in his studies directed toward the synthesis of pseudoguaianolides [187a]. The regioselectivity was explained by a directed activation involving simultaneous complexation of the ester and bridging oxygen by the TMS cation. Treatment of **216** with DBU resulted in elimination to give **217** in an overall yield of 65%, Eq. 130. The same transformation could be achieved using $BF_3\cdot Et_2O$ and KI or NaI [187b].

$$\text{215} \quad \xrightarrow[\text{MeCN, rt}]{\text{TMSI}} \quad \text{216} \quad \xrightarrow[\text{65\% yield}]{\text{DBU}} \quad \text{217} \qquad (130)$$

Vogel observed deketalization, regioselective ring opening and aromatization in one step in the reaction of **218** with TMSOTf, Eq. 131 [188]. This step was part of a sequence for the asymmetric synthesis of anthracyclinones from fused polycyclic substrates containing the [2.2.1] oxabicyclic nucleus.

$$\text{218} \quad \xrightarrow[\text{2. Ac}_2\text{O, pyr, 63\% yield}]{\text{1. TMSOTf, CH}_2\text{Cl}_2, 0°\text{C}} \qquad (131)$$

R* = RADO(Et) chiral auxiliary

Föhlisch found that treatment of 8-oxabicyclo[3.2.1]oct-6-en-3-ones with TMSOTf and triethylamine generates tropones in one step [33b]. Thus, oxabicyclic alkene **219** was converted in one step to 2-methoxytropone **220**, which is a key intermediate in several syntheses, Eq. 132 [33]. For the fully saturated derivatives **221**, the same reaction conditions produce cycloheptadienes, Eq. 133. Mann has also reported tropone formation by treatment with TMSOTf in the absence of base [189].

$$\text{219} \xrightarrow{\text{TMSOTf, Et}_3\text{N, CCl}_4,\ \text{rt}} \text{220} \tag{132}$$

$$\text{221} \xrightarrow{\text{TMSOTf, Et}_3\text{N, CCl}_4,\ \text{rt}} \tag{133}$$

4.2.4.4
Other Lewis Acids

Hoffmann found that 2,2-dialkylated 8-oxabicyclo[3.2.1]oct-6-en-3-ones such as **222** efficiently open in the presence of Lewis acid and an amine base, Eq. 134 [190]. The mechanism is apparently an enolization of the ketone followed by opening of the ether bridge. The reagent combination that was most successful was a 1:1 complex of $ZrCl_4$ and piperidine. Substrates which are not 2,2-disubstituted give tropones.

$$\text{222} \xrightarrow[\text{CH}_2\text{Cl}_2,\ -30°\text{C, 1.5 h 79\% yield}]{\text{3 equiv ZrCl}_4,\ \text{3 equiv piperidine}} \text{223} \tag{134}$$

It follows that the corresponding enol ethers can be ring-opened by treatment with Lewis acid [190]. Simpkins subjected the enantiomerically enriched silyl enol ether **224** (obtained by deprotonation using a homochiral lithium amide) to titanium tetrachloride [121]. Alkene **224** was obtained in 88% ee at $-95\,°\text{C}$, and the ring opened product is expected to be of comparable enantiomeric purity, Eq. 135.

$$\text{224} \xrightarrow[\text{76\% yield}]{\text{TiCl}_4,\ \text{CH}_2\text{Cl}_2,\ -78°\text{C}} \tag{135}$$

The acid catalyzed ring opening of 1,4-dimethyl-2,3-dicarbomethoxy-7-oxa-bicyclo[2.2.1]hepta-2,5-diene yielded the aromatized product, Eq. 136. However, in the presence of $[Rh(CO)_2Cl]_2$, methanol acts as a nucleophile and gives the cyclohexadienol. The reaction was shown to be both regio and stereoselective, Eq. 137 [191].

$$\text{(136)}$$

*catalyst, conditions = H_2SO_4, 24h, 50°C   0%   100% conversion
                        $[Rh(CO)_2Cl]_2$, 6 min, rt   60% yield   0%
                        $[Rh(CO)_2Cl]_2$, 15 h, rt   0%   75% yield

$$\text{(137)}$$

4.2.5
Grob Fragmentation

In Grieco's synthesis of compactin, the required stereochemical information in the A ring was embedded in the oxabicyclic subunit of compound **225** [192]. Ring opening was induced by base promoted Grob fragmentation which generated formaldehyde and decalin **226**, Eq. 138.

$$\text{(138)}$$

225   **226**   (+)-compactin

4.2.6
Overall Addition of Hydride

4.2.6.1
Hydrogen Addition

In the special case of the oxabicyclic compounds with bridgehead carbons bearing aryl substituents, hydrogenolysis results in the cleavage of the bridging carbon-oxygen bonds. In Rodrigo's synthesis of the lignans of *Podophyllum*, all eight diastereomers could be obtained from the common intermediate **227** [193]. In the synthesis of isopicropodophyllin, the highly-substituted cyclohexane ring in **228** was revealed by the hydrogenolysis of the oxabicyclo[2.2.1] nucleus of **227**, Eq. 139. Pelter's synthesis of (–)-isopodophyllotoxin utilized a similar hydrogenolysis strategy with an asymmetric Diels-Alder oxabicyclic adduct derived from menthol (Menth) as substrate, Eq. 140 [194].

In addition to the examples shown above, Whalley reported one case of a hydrogenation reaction that resulted in the S_N2' opening of an oxabicyclo[2.2.1] heptene [183]. In the presence of acid, substrate **229** was reductively ring opened to give **230** in quantitative yield, Eq. 141.

4.2.6.2
Single Electron Transfer Reductions

Substrates whose bridging oxygen atoms are in allylic or benzylic positions can be ring opened under dissolving metal conditions. The ring opening of oxabicyclic [4.2.1] ether **231** illustrates this reaction [43]. Treatment with lithium metal gave deprotected diol **232** as one isomer containing a tetrasubstituted olefin, Eq. 142.

(142)

231   **232**

The optimized reaction conditions for the reductive ring opening of olefinic bicyclic ether **233a** were lithium in ethylenediamine and DME [166]. A modest yield of the ring opened product **234a** was obtained due to competing simple reduction of the olefin Eq. 143. This side reaction was even more problematic for the bicyclic ether **233b**, in which the desired reductive ring opening gave (+)-dactylol **234b** in lower yield than the side-product, **235b**.

(143)

233a R= H
233b R= Me

234a   45%   **235a** 27%
234b (+)-dactylol 25%   **235b** 35%

In addition to ring opening, the reaction of sodium with the oxabicyclic substrate **236** resulted in elimination of methoxymethoxide and reduction of the diene [118]. Only one olefinic product **237** was isolated, Eq. 144.

(144)

236   **237**

The reductive ring opening can also be induced by single electron donor reagents. In De Clercq's formal total synthesis of periplanone B, oxabicyclic intermediate **238** was reductively ring opened by treatment with lithium di-*tert*-butylbiphenyl radical anion, Eq. 145 [195]. Subsequent Grob fragmentation leading to scission of the ring junction bond generated the decadienone **239** which has been transformed into periplanone B.

(145)

238   **239**   (±)-periplanone B

William's synthesis of a model compound of the dolastanes employed sodium naphthalide to induce the ring opening of the allylic ether, Eq. 146 [59]. Protonation at the γ carbon gave the conjugated enone.

$$(146)$$

Carbonyl groups alpha to the bridging oxygen undergo reduction in the presence of samarium iodide, resulting in ketyl radical anion formation and fragmentation of the carbon-oxygen bond. This reductive ring opening was used by Padwa in synthetic studies toward ptaquilosin [72]. Treatment of **240** with SmI$_2$ generated **241** which contains the basic skeleton of the target molecule. It is noteworthy that the cyclopropyl substituent remained intact under the reaction conditions, Eq. 147.

$$(147)$$

240   **241**   Ptaquilosin

The samarium iodide promoted reduction of substrate **242** also led to ring opening to yield hydroxycyclohexenone **243** in De Clercq's synthesis of a precursor to the A-ring of 1 α-hydroxyvitamin D$_3$, Eq. 148 [196].

$$(148)$$

242   **243**

1α, 25-(OH)$_2$ Vitamin D$_3$

In Vogel's studies, the [2.2.1] oxabicyclic substrate **244** was found to undergo reductive ring opening as well as thermodynamic protonation to furnish a cyclohexanol, Eq. 149 [197].

$$ (149) $$

Enantiomerically enriched substrate **245** was found to undergo reductive ring-opening in the presence of SmI_2; however, much more efficient opening was observed using lithium in ammonia, Eq. 150 [120].

$$ (150) $$

4.2.6.3
Reductive Elimination

Halides and sulfones positioned at the carbon alpha to the bridging ether bond can be induced to undergo reductive elimination leading to ring opening. Jung's model studies toward the synthesis of ivermectin utilized this strategy [198]. The key substrate **246** was assembled by intramolecular Diels-Alder reaction of an *N*-furfuryl-β-chloroacrylamide followed by dihydroxylation. Treatment with sodium resulted in ring opening to afford the bicyclic trihydroxy amide **247** found in the "southern hemisphere" of ivermectin, Eq. 151.

$$ (151) $$

Sammes synthesized β-bulnesene by employing a sodium reduction of chloroether **248** to effect the ring opening of the bridging C–O bond in a [3.2.1] oxabicyclic system, Eq. 152 [56].

$$ (152) $$

Wender incorporated this strategy into the synthetic plan for the first total synthesis of phorbol, whereby intermediate **249** was subjected to lithium-iodine exchange to yield alkenol **250**, Eq. 153 [199].

$$(153)$$

A recent example of a ring opening based on the same principle is found in a series of synthetic studies toward taxol, in which model compound **251** has an oxabicyclo[2.2.1]heptane moiety derived from furfuryl alcohol as the precursor for ring-C of the target [200]. The hydroxymethyl group in **251** was converted to the iodide, and treatment with freshly activated zinc resulted in ring opening to the tricyclic system **252**, Eq. 154.

$$(154)$$

Samarium iodide has been used to reduce sulfonylated oxabicyclic substrates leading to the elimination of the β oxygen moiety. Molander used this strategy for the synthesis of substituted cycloheptenes and cyclooctenes, Eq. 155 [81].

$$(155)$$

4.2.6.4
Metal Hydride Reductions

4.2.6.4.1
β-Hydridic Organometallic Reagents

Grignard reagents react sluggishly with oxabicyclic compounds in the absence of a transition metal catalyst. In the presence of excess $MgBr_2$, the products of reductive ring opening predominate [201]. Therefore in the presence of *n*-butyl-lithium and excess $MgBr_2$ (which forms *n*-butylmagnesium bromide), oxabicyclic substrate **253a** gives cyclohexenol **255**, Eq. 156.

$$\text{(156)}$$

The reductive ring opening could be explained by a mechanism with a transition state resembling **254**, in which the β-hydrogens of the Grignard reagent reduce the double bond. The mechanism accounts for the requirement of additional $MgBr_2$, and also suggests that the structure and the number of β-hydrogens of the Grignard reagent should have an effect on the reductive ring opening. Indeed, variations in regioselectivity were observed when different Grignard reagents were used in reductive ring openings of unsymmetrical substrate **256a**, Eq. 157. However, the low yields and selectivities make this reaction of mechanistic interest rather than of practical value.

$$\text{(157)}$$

R= t-Bu	Yields: 50%	1.2	:	1
R= n-Bu	37%	1	:	2.4
R= i-Bu	44%	1	:	4.5

The product from reductive ring opening was isolated along with the product from the nucleophilic addition in the reaction of $t\text{-Bu}_2\text{CuCNLi}_2$ with oxabicyclic substrate **253b**, Eq. 158, vide infra [202]. Reduction by of one of the β-hydrogens of the *tert*-butyl group of the cuprate must be responsible for this product.

$$\text{(158)}$$

4.2.6.4.2
Boranes and Borohydrides

Brown reported that the reagent used for the reductive cleavage of cyclic ethers, a lithium triethylborohydride-aluminum *tert*-butoxide complex (from lithium tri(*tert*-butoxy)aluminum hydride and triethylborane), when applied to 7-oxabicyclo[2.2.1]heptane, gave cyclohexanol, Eq. 159 [203].

$$\text{LiAl(Ot-Bu)}_3\text{H / Et}_3\text{B, THP, rt, 3h} \qquad \text{97\% yield} \tag{159}$$

In the context of dihydronaphthalene oxides, Rickborn showed that a related complex induced ring-opening of **257** with S_N2 delivery of hydride, Eq. 160 [204].

$$\text{LiAl(Ot-Bu)}_3\text{H / B(OEt)}_3\text{, THP} \tag{160}$$

257

This result is in contrast to the reaction of **257** with less sterically demanding hydroborating reagents such as borane and 9-BBN, which delivers the hydride in an S_N2' fashion to yield a homoallylic alcohol, Eq. 161 [97]. The mechanism was proposed to be addition of borane followed by a *syn*-elimination aided by chelation to the bridging oxygen. This proposal accounts for the observation that bulky boranes, such as Sia$_2$BH, led only to simple hydroboration products without inducing ring cleavage, since the alkylborane was too hindered to coordinate to the bridging ether.

$$\begin{array}{l} 1.\ \text{BH}_3\text{·Me}_2\text{S ,THF, }\Delta\text{ or 9-BBN , THF, rt} \\ 2.\ \text{NaOH, H}_2\text{O}_2 \end{array} \tag{161}$$

257   quantitative yield

4.2.6.4.3
Aluminum Hydrides

Various aluminum hydrides have been found to induce the reductive ring opening of [2.2.1] and [3.2.1] oxabicyclic compounds. Metz found that the treatment of sultone **258** with Red-Al resulted in the overall net S_N2' addition of hydride and ring opening [165]. When **260** was found to also give **261** under the same reaction conditions, the mechanism postulated to account for this transformation was an initial deprotonation of the sultone **258** by Red-Al and ring opening, followed by the 1,6-delivery of hydride via aluminate **259**, and stereoselective protonation, Eq. 162.

$$(162)$$

Another example of reductive ring opening of oxabicyclic substrates was provided by Arjona and co-workers [117a]. LAH induced the ring opening of sulfonylated oxabicyclic compounds, and the regioselectivity of the addition was dictated by the position of the vinyl sulfone moiety, Eq. 163.

$$(163)$$

Vogel also reported reductive ring opening in substrates containing the vinyl sulfone functionality in syntheses of acyclic subunits containing four contiguous stereocenters, Eq. 164 [109b].

$$(164)$$

Lautens and Chiu showed DIBAL-H was a useful reagent for the efficient reductive ring opening of a wide range of oxabicyclo[2.2.1] and [3.2.1] substrates [205]. The efficiency of DIBAL-H in ring opening reactions is attributed to its solubility, reducing ability and Lewis acidity, which enable it to coordinate to the ether bridge and facilitate the cleavage step. S_N2' delivery of hydride generates homoallylic alcohols such as 263 from 262, Eq. 165. Subsequent manipulations of 263 including ring cleavage by ozonolysis afforded the terminally differentiated array 265, which is the C_{17} to C_{23} subunit of ionomycin, Eq. 166, Fig. 7.

$$(165)$$

| Yields: | 25% recovered **262a** | 65% | 0% |
| | 0% recovered **262a** | 50% | 27% |

$$(166)$$

Ionomycin

Fig. 7

In the case of [3.2.1] oxabicyclic substrates unsymmetrically substituted at the bridgehead position, an interesting regioselectivity was noted, Eq. 167. With **266a** protected as a *tert*-butyldimethylsilyl ether, hydride delivery proximal to the hindered bridgehead was favored. No selectivity was observed in the reductive ring opening of the free alcohol **266b**. However, treatment with MeLi then DIBAL-H (i.e. **266c**) results in a dramatic reversal of regioselectivity compared to the protected ether **266a**.

$$(167)$$

266a	R= TBDMS	Yields: 72%	1	:	6.4
266b	R= H	75%	1	:	1.3
266c	R= Li	85%	9.5	:	1

Keay also reported an example of a DIBAL-H promoted reductive ring opening. While several similar substrates were not reactive with DIBAL-H, the cycloadduct **267** underwent reductive ring opening along with carbocyclic ring cleavage, Eq. 168 [206].

$$\text{(168)}$$

267   28%   13%

One problem associated with the use of DIBAL-H under such vigorous conditions (i.e. refluxing hexanes) is the appearance of an over-reduced side product, **264**, which was difficult to separate from the desired cycloheptenol **263**, Eq. 165. The presence of this product indicates a lack of chemoselectivity associated with DIBAL-H for some types of cyclic alkenes versus an olefin in an oxabicyclic system.

A milder and more selective reductive ring opening was achieved in a nickel-catalyzed hydroalumination [207]. The initial addition of DIBAL-H to the oxabicyclic alkene under Ni^0-catalysis occurs at temperatures as low as $-78\,°C$, and is complete in minutes at room temperature. Oxabicyclo[2.2.1] substrates such as **253c** spontaneously undergo ring opening under the reaction conditions, Eq. 169. The less strained [3.2.1] oxabicyclic compounds require heating of the organoalane in the presence of DIBAL-Cl to induce ring opening. The two-step, one pot sequence led to substantially improved yields of the desired ring opened product, accompanied by less than 5% over-reduction. A particularly dramatic example of the efficiency of the nickel-catalyzed reduction is illustrated in Eq. 170. Treatment of **268** with DIBAL-H in the absence of a catalyst gives a 1:1 ratio of **269** and *cis* 1,3-cycloheptanediol. Using nickel catalysis followed by a Lewis acid, a 95:5 ratio favoring **269** was obtained. *meso* Diol **269** has been used as a precursor in a concise and enantioselective synthesis of the mevinic acid lactone, the portion of mevinolin to which its biological activity largely resides [113].

$$\text{(169)}$$

253c   3–10 mol $Ni(COD)_2$ DIBAL-H, THF or toluene   >90%

$$\text{(170)}$$

268   1. cat $Ni(COD)_2$, DIBAL-H   2. DIBAL-Cl, Δ   85% yield   **269**   mevinic acid lactone

Transition metal-catalyzed reductive opening also allowed the use of coordinating ligands to tune the reactivity of the reagent. Two significant findings have resulted from these studies. The addition of 1,4-bis(diphenylphosphino)butane (dppb) dramatically enhanced the regioselectivity of the reductive ring opening of substrates unsymmetrically substituted at the bridgehead position [207, 208]. For example, the nickel-catalyzed reductive ring opening of **266a** generated **270** and **271** in a 2.1:1 ratio, Eq. 171. Addition of dppb increased the regioselectivity of hydride delivery distal to the bridgehead methyl group more than twenty-fold (98:2).

Another important development has been the use of chiral phosphines as ligands to induce an enantioselective reductive ring-opening of *meso* oxabicyclic compounds. BINAP, available in both (*R*)- and (*S*)-forms, gives the highest enantioselectivities of the ligands examined to date with values of 97% ee for the [2.2.1] oxabicyclic substrate **253c** under the optimized conditions, Eq. 172 [207].

4.2.6.4.4
Tin Hydrides

Lautens and Klute reported a regioselective palladium-catalyzed hydrostannylation of oxabicyclic substrates bearing substituents at the bridgehead position [209a]. A variety of oxabicyclo[2.2.1] compounds such as **256b** undergo regioselective addition of tin hydride such that the bulky trialkyltin resides at the less hindered position, Eq. 173. The regioselectivity is generally at least 97:3.

The stannylated product **272** can be induced to undergo ring-opening by treatment with MeLi, either via a transmetallation or the ate complex. This overall sequence provides the reductive ring opened product **273** with complementary regioselectivity to that obtained through nickel and phosphine-catalyzed hydroalumination.

Oxabicyclic compounds in the [3.2.1] series also undergo highly regioselective hydrostannation and MeLi induced ring opening under these conditions. However, more hindered alkenes are efficiently hydrostannated using a heterogeneous palladium catalyst [209b]. In this manner, cycloheptenetriol **275** was produced from **274**, the product of the diastereoselective [4+3] cycloaddition (Scheme 12) [45]. The chiral side chain was cleaved by periodate oxidation. Reduction afforded diol **276**, an intermediate which has been used previously for a synthesis of the C_{17} to C_{23} subunit of ionomycin (see Eq. 166). This route constitutes an enantioselective synthesis of this stereochemical array.

Scheme 12

4.2.6.5
Photochemical Reductions

Cossy has shown that strained cyclopropanes and cyclobutanes situated alpha to a carbonyl group open via the ketyl radical anions formed during photolyses in the presence of amines [197, 210]. Moreover, strained ethers such as oxygen bridged bicyclic compounds have also been observed to undergo opening under these conditions, as shown by Eq. 174.

(174)

4.2.7
Overall Addition of Alkyl/Aryl Groups

4.2.7.1
Silyl Enol Ether and a Lewis Acid

Narasaka found that optically enriched oxabicyclic substrate 277 bearing a vinyl sulfide moiety reacts with a silyl enol ether or ketene silyl acetal in the presence of a Lewis acid to afford the protected cyclohexenols 278a and 278b, Eq. 175 [18]. The reaction was proposed to occur via a ring-opening and alkylation sequence which is equivalent to overall nucleophilic substitution with retention of configuration. Presumably, the nucleophile attacked the carbocationic intermediate from the *exo* face, because the methylene-OTIPS substituent was blocking the *endo* side.

278a R= Ph, 89% yield
278b R= OEt, 91% yield

(175)

4.2.7.2
Organolithium Reagents

The earliest reports of the addition of organolithium reagents to oxabicyclic compounds were in the context of dihydronaphthalene oxide 257. Caple and Berchtold found that the additions occur in an S_N2' fashion, leading to alcohols 279a–c, Eq. 176 [211, 212].

279a R= n-Bu, 85% yield
279b R= t-Bu, 100% yield
279c R= Me, 61% yield

(176)

The groups of Arjona and Lautens independently investigated the addition of organolithium reagents to oxabicyclic substrates. Arjona discovered that treatment of oxabicyclo[2.2.1]heptenol 280a, readily prepared using Vogel's naked sugar chemistry, with an excess of an organolithium reagent, resulted in ring opening [105]. The reaction was completely regioselective and stereoselective; for example cyclohexenediol 281a was isolated in good yield, Eq. 177. Because

$$n\text{-BuLi, Et}_2\text{O} \quad\quad\quad\quad (177)$$

280a R^1= Me, R^2= OH
280b R^1= OH, R^2= Me

281a 78% yield
281b 75% yield

280a is available in enantiomerically pure form, the ring opened product is a single enantiomer. The *exo* alcohol **280b** also underwent regioselective nucleophilic ring opening, although more vigorous reaction conditions were required. The reason for the directing effect of the lithio alkoxide has not been elucidated.

However, when the hydroxyl group was protected as a benzyl ether, the regioselectivity decreased dramatically, Eq. 178 [105], as it did for the homologous alcohol, Eq. 179.

$$3 \text{ equiv. } n\text{-BuLi}, \quad \text{Et}_2\text{O, 0°C} \quad\quad\quad (178)$$

2 : 1

$$n\text{-BuLi}, \quad \text{Et}_2\text{O, 0°C}, \quad 80\% \quad\quad\quad (179)$$

3.5 : 1

The directing effect of a hydroxyl group alpha to the bridgehead carbon was also observed with an oxabicyclo[3.2.1] substrate, although only *t*-BuLi is sufficiently reactive to induce ring opening, Eq. 180 [213].

$$\text{excess } t\text{-BuLi, Et}_2\text{O, 0°C to rt}, \quad 84\% \text{ yield} \quad\quad\quad (180)$$

Sulfonylated derivatives **282** and **283** were designed to show how an electron-withdrawing group could direct the regioselectivity of the addition of the organolithium reagent [117, 213, 214]. Because both regioisomers could be synthesized, this aim was realized as shown by Eqs. 181, 182. Methyllithium-induced opening of a related substrate, **284**, was used in a synthesis of the amino-cyclitol portion of pancratistatin, Eq. 183 [215].

$$(181)$$

$$(182)$$

$$(183)$$

aminocyclitol of pancratistatin

Concurrent with Arjona's work were studies by Lautens and co-workers on the organolithium induced ring opening of oxabicyclo[3.2.1]octenes **262b–c** [112]. All organolithium reagents that successfully induce ring opening give products which can be rationalized by an S_N2' reaction with retention of stereochemistry. Cycloheptenyl homoallylic alcohols **285a–b** are readily available, Eq. 184. The reactivity of the nucleophiles correlates with the basicity (and/or electron transfer ability) of the organolithium reagents.

$$(184)$$

262b R= H
262c R= TBDMS

285a 92% yield
285b 79% yield

The ring opening of oxabicyclo[3.2.1]octenol **262b** was more facile than its protected counterpart **262c** [116]. This enhancement of reactivity by the remote *endo*-alkoxide was most dramatically displayed in the nucleophilic ring opening by MeLi, Eq. 185. Under the typical reaction conditions, **262b** resisted ring opening, due to the low nucleophilicity of MeLi in this reaction. Addition of TMEDA was necessary to bring about opening, affording a 72% yield of **286a**. However, **262c** was totally inert even when heated in TMEDA. The two hydroxyl groups of **286a** were sequentially protected to give **286b** and the cleavage of the olefin eventually led to the synthesis of the C_{21}–C_{27} subunit of rifamycin, Eq. 185.

The regiochemistry of the reaction was examined in the context of unsymmetrical oxabicyclic substrates bearing a substituent at the bridgehead position [216]. An ethyl group, which is not very sterically demanding, induced highly regioselective ring-opening reactions in which the nucleophile was delivered to

(185)

the position distal to the bridgehead substituent, Eq. 186. The cycloheptenol **287** thus obtained from **266b** was subjected to ozonolysis to furnish an acyclic chain bearing 5 contiguous stereocenters with differentiated termini. The high regioselectivity may indicate that complexation of lithium to the bridging oxygen weakens the C–O bond to generate the more stable cation. Delivery of the nucleophile then occurs remote to the bridgehead substituent.

(186)

The increased strain in oxabicyclo[2.2.1]heptenes such as **256b** makes them more reactive toward organolithium reagents. The ring opening reactions occurred at lower temperatures and with higher regioselectivities, Eq. 187.

(187)

Keay and Harmata obtained additional information on these trends in polycyclic oxygen bridged compounds. Thus treatment of **288** with MeLi led to ring opening with the addition of the nucleophile distal to the bridgehead to give **289**, Eq. 188 [206]. Subsequent manipulations of **289** led to a synthesis of the C_{15}–C_{23} segment of the venturicidins, Fig. 8. Reaction of cycloadduct **290** with isopropyllithium revealed a slight preference for the position near the methyl-substituted bridgehead over the ring junction, Eq. 189. Regioselective ring opening also occurred for the reaction of polycyclic substrate **291** with *n*-butyllithium, Eq. 190 [42].

(188)

(189)

(190)

venturicidin X

Fig. 8

Dioxacyclic compounds have also been shown by Lautens and Fillion to undergo regio-, stereoselective and sequential ring opening, Scheme 13 [216]. Whereas reaction of the dioxacyclic compound at 0 °C led to incorporation of

Scheme 13

two *n*-butyl groups, reaction at −78 °C was very selective and could be stopped after only one oxabicyclic moiety underwent opening. Addition of a second, different nucleophile could then be readily achieved providing a route to highly functionalized decalins.

Metz showed that unsymmetrical sultones undergo regio- and stereoselective alkylative ring opening via elimination/1,6-addition when treated with organolithium reagents [217a]. The stabilized carbanion from attack of the nucleophile *syn* to the intermediate alkoxide can be trapped by acid or MeI, Eq. 191. Desulfurization led to functionalized cyclohexenols with stereochemical control on the ring as well as the side chain. Such a ring opening has been used in a short synthesis of methyl nonactate [217b].

$$\text{(191)}$$

Lautens explored the behavior of silyllithium reagents with [2.2.1] oxabicyclic substrates, typified by **253a**, and **256b** [218]. Cyclohexadienes **292** and **293** respectively were isolated as the only products in good yields, Eq. 192. The reaction was proposed to occur via a nucleophilic ring-opening by silyllithium, which generated an intermediate with the alkoxide and silyl substituent in a *syn* relationship. A Peterson elimination occurred spontaneously under the basic reaction conditions and gave rise to the conjugated dienes. Therefore, the silyllithium reagent provides a one-step synthesis of cyclohexadienes from oxabicyclic precursors. The intermediate hydroxysilane was isolated in one case providing further support for this mechanistic proposal.

$$\text{(192)}$$

253a R=H, R'=TBDMS
256b R= Me, R'=TIPS

292 R=H, R'=TBDMS
293 R= Me, R'=TIPS

Although the organolithium-induced ring opening of [2.2.1] oxabicyclic substrates has been reported to occur in DME, a dramatic effect was observed in the corresponding reaction of oxabicyclo[3.2.1]octenes. Cycloheptadienes such as **294**, the product of addition-ring opening and dehydration of **262b**, were obtained under the otherwise typical nucleophilic ring opening conditions, Eq. 193 [219]. This reaction pathway was not observed when the hydroxyl group was protected, once again pointing to an usual *endo*-alkoxide effect.

An intramolecular nucleophilic addition to construct fused bicyclic systems was recently developed by Lautens and Kumanovic [220]. The iodopropyl-substituted oxabicyclic substrate **295a** underwent transmetallation with *t*-BuLi

$$\text{262b} \xrightarrow[\text{0°C, 4 h, 85\%}]{\text{MeLi, DME}} \text{294} \tag{193}$$

at $-78\,°C$, and upon warming to room temperature, intramolecular addition and ring opening occurred to give **296a** in high yield, Eq. 194. It is particularly significant that this reaction generated a *trans* junction in the perhydroazulene skeleton, which is the stereochemistry found in natural products such as phorbol, daphnane and grayanotoxin. Heteroatoms in the tether were also tolerated, the precursors in these cases being stannylated oxabicyclo[3.2.1] compounds **295b–d**. A four-atom tethered substrate failed to undergo intramolecular opening.

$$\xrightarrow[\text{-78°C to 0°C or rt}]{t\text{-BuLi or MeLi, -78°C}} \tag{194}$$

295a R= Me, X= CH$_2$, Z= I
295b R= Me, X= O, Z= SnBu$_3$
295c R= Me, X= S, Z= SnBu$_3$
295d R= H, X= NMe, Z= SnBu$_3$

296a 80%
296b 82%
296c 80%
296d 80%

Asymmetric induction in the organolithium ring opening of *meso* oxabicyclic compounds was achieved by incorporating a catalytic amount of sparteine as an additive, Eq. 195 [221]. Sparteine increased the reactivity of the organolithium reagent toward ring opening as well as induced modest enantioselectivity ($\leq 52\%$ ee) in the reaction.

$$\xrightarrow[\substack{\text{pentane, -40°C} \\ \text{52\% ee, 60\% yield}}]{\text{15 mol\% sparteine, } n\text{-BuLi}} \tag{195}$$

4.2.7.3
Organocuprate Reagents

Lautens examined the reaction of cuprates with [3.2.1] oxabicyclic substrate **297** and found that the major reaction pathway is an S_N2' addition-ring opening, but contrary to the usual *syn* opening, an anti addition of the nucleophile was observed. Minor products due to anti-S_N2 addition to the olefin and addition to the carbonyl group were also obtained, Eq. 196 [222].

$$(196)$$

R= Me	Yields: 66%	15	:	<1	:	1
R= *n*-Bu	59%	15	:	<1	:	1
R= *s*-Bu	85%	8	:	1	:	<1
R= *t*-Bu	92%	3	:	1	:	<1

With oxabicyclo[2.2.1]heptenes, S_N2' addition *syn* to the oxygen bridge occurred exclusively to give good yields of the homoallylic cyclohexenol, Eq. 197 [202].

$$(197)$$

R= *t*-Bu   Yields: 78%
R= *s*-Bu   85%

Higher order cuprates also ring opened unsymmetrical oxabicyclo[2.2.1] heptene **298** with good regioselectivity, Eq. 198; however, no selectivity was observed in the reaction with unsymmetrical substrates such as **299**, Eq. 199 [202].

$$(198)$$

$$(199)$$

1   :   1

The reactivity of silylcuprates with oxabicyclic compounds was also examined [218, 223]. With oxabicyclo[2.2.1] compounds, addition and Peterson elimination to produce cyclohexadienes occurred as with silyllithium reagents, Eq. 200.

However, with oxabicyclo[3.2.1] compounds, the product from addition to the olefin and trapping by the ketone were detected rather than the typical ring opening reaction, Eq. 201.

$$\text{(200)}$$

253a R^1, R^2=H, R^3=TBDMS
256a R^1, R^2=Me, R^3=TBDMS
253b R^1=H, R^2=Me, R^3=TIPS
256c R^1=H, R^2=Me, R^3=Bn

292 73%
293b 83%
293a 78%
293c 81%

$$\text{(201)}$$

1. $(PhMe_2Si)_2CuCNLi_2$, 0°C, THF

2. H_2O, 95% yield

297

2.2 : 1

4.2.7.4
Transition Metal-Catalyzed Alkylative Ring-Opening

Cheng recently reported the palladium-catalyzed addition of iodoarenes and alkenes to 7-oxabenzonorbornadiene derivatives which resulted in overall alkylation and ring opening, affording products **300a–d**, Eq. 202 [224]. This methodology complements the existing organolithium induced ring openings because the corresponding lithioarenes and alkenes are typically poor nucleophiles for this process.

$$\text{(202)}$$

RI, $Pd(PPh_3)_2Cl_2$

Zn, $ZnCl_2$, Et_3N, THF, Δ

257

300a R=Ph, 86% yield
300b R=Bn, 99% yield
300c R= 2-thiophene, 71% yield
300d R= , 83% yield

Asymmetric induction by the use of chiral phosphines was explored in the palladium-catalyzed phenylation of **257** [225]. The yields and enantioselectivities of the ring opened products are highly variable. For example, **300a** was obtained in 96% ee with (*R*)-BINAP as ligand but the yield was very poor, Eq. 203. The addition of $ZnCl_2$ increased the yield of the ring opened product to

$$\text{(203)}$$

Ph(OTf), Pd, *(R)*-BINAP

NaOOCH, DMF

257

300a R=Ph, 13% yield, 96% ee

41 % but the enantioselectivity was significantly diminished (54 % ee). Using PhI instead of the triflate gave a racemic product.

Another potentially useful ring opening reaction which complements the existing methodology was realized in the nickel-catalyzed addition of Grignard reagents to oxabicyclic substrates [226]. Ring opening by a methyl or phenyl nucleophile was achieved, which were unreactive in the absence of catalyst as were the analogous organolithium reagents. Substrates such as **256c** bearing a bridgehead substituent, Eq. 204, or **262d** in which the *endo* hydroxyl group was protected, gave the previously unavailable products, Eq. 205. Interestingly, the use of Ni(dppp)Cl$_2$ as catalyst with HMPA as co-solvent led to products in which the nucleophile added *trans* with respect to the oxygen bridge (Scheme 14). Formation of an alkyl-π-allyl nickel complex, and reductive elimination may be responsible for the stereochemical outcome of the reaction.

$$\textbf{256c} \xrightarrow{\text{cat Ni(COD)}_2,\ \text{MeMgBr, Et}_2\text{O reflux}} \text{74\% yield} \qquad (204)$$

$$\textbf{262d} \xrightarrow[\text{56\% yield}]{\text{cat Ni(COD)}_2,\ \text{MeMgBr, Et}_2\text{O reflux}} \qquad (205)$$

$$\textbf{253c} \xrightarrow[\substack{\text{Et}_2\text{O, HMPA, rt} \\ \text{95\% yield}}]{\substack{\text{cat Ni(dppp)Cl}_2 \\ \text{MeMgBr}}} \quad \text{or} \quad \longrightarrow \quad 1.4{:}1 \text{ to } 1{:}2.4$$

Scheme 14

5
Conclusions and New Frontiers

New stereoselective chemical reactions and new strategies for the synthesis of stereochemically complex bioactive compounds remains a focus of intense activity in organic chemistry. In this review, we have shown that oxabicyclic compounds have become valuable intermediates which can address these needs because of the high stereocontrol observed in many of the reactions of these rigid molecules. However, improved methods of synthesis of symmetrical and unsymmetrical compounds are required as are a wider range of enantioselective transformations. Reliable methods for the synthesis of larger bicyclic ethers are needed so that cyclooctanyl and larger rings can be prepared.

The enantioselective opening of *meso* compounds is a highly efficient entry to optically active cyclic and acyclic compounds and is an area awaiting further breakthroughs. Transition-metal catalyzed processes may lead to milder and more selective reactions on increasingly complex substrates.

The full impact and applicability of the ring opening strategy will not be fully delineated for some time.

Acknowledgements. We thank the E. W. R. Steacie Foundation, NSERC Canada, the Alfred P. Sloan Foundation, the Merck Frosst Centre for Therapeutic Research, BioMéga/Boehringer Ingelheim, Allelix Biopharmaceuticals, Eli Lilly, Pharmacia/Upjohn and the University of Toronto for their support of our programs. We thank Tomislav Rovis and Renée Aspiotis for their helpful comments on an early draft of the manuscript. P.C. thanks Professor K.F. Cheng of the University of Hong Kong for his support during the writing of this review.

References

1. Lipshutz BH (1986) Chem Rev 86:795
2. (a) Moore JA, Partain EM (1983) J Org Chem 48:1105 (b) Nugent WA, McKinney RJ, Harlow RL (1984) Organometallics 3:1315 (c) Bailey MS, Brisdon BJ, Brown DW, Stark KM (1983) Tetrahedron Lett 24:3037
3. Moursoundidis J, Wege D (1983) Aust J Chem 36:2473
4. Laszlo P (1986) Acc Chem Res 19:121
5. Ipaktschi J (1986) Z Naturforsch Teil B 41:496
6. Saksena AK, Girijavallabhan VM, Chen Y-T, Jao E, Pike RE, Desai JA, Rane D, Ganguly AK (1993) Heterocycles 35:129
7. Dolata DP, Bergman R (1987) Tetrahedron Lett 28:707
8. Matsumoto K, Sera A (1985) Synthesis 999
9. Sargent MV, Dean FM (1984) Furans and their Benzo Derivatives, (ii) Reactivity. In: Katritzky AR, Rees CW (eds) Comprehensive Heterocyclic Chemistry. vol 4, Pergamon Press, Oxford, p 599
10. Shipman M (1995) Contemporary Org Synth 2:1
11. (a) Vogel P, Fattori D, Gasparini F, Le Drian C (1990) Synlett 173 (b) Reymond JL, Vogel P (1990) Asymmetry 1:729 (c) Vogel P (1990) Bull Soc Chim Belg 99:395 (d) Carrupt PA, Vogel P (1984) Tetrahedron Lett 25:2879 (e) Carrupt PA, Vogel P (1988) J Phys Org Chem 1:287
12. Black KA, Vogel P (1984) Helv Chim Acta 67:1612
13. Vieira E, Vogel P (1983) Helv Chim Acta 66:1865
14. Kernen P, Vogel P (1993) Tetrahedron Lett 34:2473
15. Takayama H, Iyobe A, Koizumi T (1986) J Chem Soc Chem Commun 771
16. Takahashi T, Namiki T, Takeuchi Y, Koizumi T (1988) Chem Pharm Bull 36:3213
17. (a) Corey EJ, Loh TP (1993) Tetrahedron Lett 34:3979 (b) Evans has recently shown that a cationic bis (oxazoline) Cu(II) catalyst is effective for the enantioselective cycloaddition between furan and acryloyl oxazolidinone, Evans DA, Barnes DM (1997) Tetrahedron Lett 38:57
18. Yamamoto I, Narasaka K (1995) Chem Lett 1129
19. Roush WR (1990) Stereochemical and Synthetic Studies of the Intramolecular Diels-Alder Reaction. In: Curran DP (ed) Advances in Cycloaddition. vol 2, JAI, Greenwich, p 91

20. (a) Craig D (1987) Chem Soc Rev 16:187 (b) Taber DF 'Intramolecular Diels-Alder and Alder Ene Reactions' Springer Berlin 1984 (c) Fallis AG (1984) Can J Chem 62:183 (d) Ciganek E (1984) Org React 32:1
21. Fischer K, Hünig S (1987) J Org Chem 52:564
22. Bovenschulte E, Metz P, Henkel G (1989) Angew Chem Int Ed Engl 28:202
23. Woo S, Keay BA (1994) Tetrahedron Asymm 5:1411
24. Mann J (1986) Tetrahedron 42:4611
25. Reviews on the [4+3] cycloaddition of oxyallyl cations: (a) Rigby JH, Pigge FC to appear in Organic Reactions vol 51 (b) Hosomi A, Tominaga Y (1991) [4+3] Cycloadditions. In: Trost BM, Fleming I (eds) Comprehensive Organic Synthesis. vol 5, Pergamon, Oxford, p 593 (c) Mann J (1986) Tetrahedron 42:4611 (d) Noyori R, Hayakawa Y (1983) Org React 29:163 (e) Harmata M to appear in Lautens M (ed) Advances in Cycloadditions, vol 4, JAI Press and also ref. 128
26. (a) Hoffmann HMR (1973) Angew Chem Int Ed Engl 12:819 (b) Hoffmann HMR; Clemens KE; Smithers RH (1972) J Am Chem Soc 94:3940
27. Mann J, Barbosa LCA (1992) J Chem Soc Perkin Trans 1 787
28. Fukuzawa S, Fukushima M, Fujinami T, Sakai S (1989) Bull Chem Soc Jpn 62:2348
29. (a) Joshi NN, Hoffmann HMR (1986) Tetrahedron Lett 27:687 (b) Hoffmann HMR, Eggert U, Gibbels U, Giesel K, Koch O, Lies R, Rabe J (1988) Tetrahedron 44:3899
30. Herter R, Föhlisch B (1982) Synthesis 976
31. Takaya H, Makino S, Hayakawa Y, Noyori R (1978) J Am Chem Soc 100:1765
32. Vinter JG, Hoffmann HMR (1973) J Am Chem Soc 95:3051
33. (a) Murray DH, Albizati KF (1990) Tetrahedron Lett 31:4109 (b) Föhlisch B, Sendelbach S, Bauer H (1987) Liebig Ann Chem 1
34. (a) Föhlisch B, Krimmer D, Gerlach E, Käshammer D (1988) Chem Ber 121:1585 (b) Föhlisch B, Herter R, Wolf E, Stezowski JJ (1982) Chem Ber 115:355
35. Sasaki T, Ishibashi Y, Ohno M (1982) Tetrahedron Lett 23:1693
36. Ohno M, Mori K, Hattori T, Eguchi S (1990) J Org Chem 55:6086
37. Erden I, Amputch MA (1987) Tetrahedron Lett 28:3779
38. Noyori R, Nishizawa M, Shimizu F, Hayakawa Y, Maruoka K, Hashimoto S, Yamamoto H, Nozaki H (1979) J Am Chem Soc 101:220
39. Föhlisch B, Herter R (1984) Chem Ber 117:2580
40. (a) Harmata M, Gamlath CB (1988) J Org Chem 53:6154 (b) Harmata M, Gamlath CB, Barnes CL (1990) Tetrahedron Lett 31:5981
41. Harmata M, Fletcher VR, Claassen II RJ (1991) J Am Chem Soc 113:9861
42. (a) Harmata M, Elahmad S (1993) Tetrahedron Lett 34:789 (b) Harmata M, Jones DE (1996) Tetrahedron Lett 37:783
43. (a) Harmata M, Elahmad S, Barnes CL (1994) J Org Chem 59:1241 (b) Harmata M, Elomari S, Barnes CL (1996) J Am Chem Soc 118:2860
44. Schultz AG, Macielag M, Plummer M (1988) J Org Chem 53:391
45. Lautens M, Aspiotis R, Colucci JT (1996) J Am Chem Soc 118:10930
46. Henning R, Hoffmann HMR (1982) Tetrahedron Lett 23:2305
47. (a) Harmata M, Herron BF (1993) J Org Chem 58:7393 (b) Harmata M, Jones DE (1997) J Org Chem 62:1578
48. Harmata M, Herron BF (1993) Tetrahedron Lett 34:5381
49. West FG, Chase CE, Arif AM (1993) J Org Chem 58:3794
50. (a) Davies HML, Clark DM, Smith TK (1985) Tetrahedron Lett 26:5659 (b) Davies HML, Clark DM, Alligood DB, Elband GR (1987) Tetrahedron 43:4265 (c) Davies HML, Ahmed G, Churchill MR (1996) J Am Chem Soc 118:10774
51. Sammes PG (1986) Gazz Chim Ital 119:109
52. Katritzky AR, Dennis N (1989) Chem Rev 89:827
53. Hendrickson JB, Farina JS (1980) J Org Chem 45:3359
54. Sammes PG, Street LJ (1983) J Chem Soc Perkin Trans I 1261
55. Sammes PG, Street LJ (1982) J Chem Soc Chem Commun 1056
56. Bromidge SM, Sammes PG, Street LJ (1985) J Chem Soc Perkin Trans I 1725

57. (a) Sammes PG, Street LJ (1983) J Chem Soc Chem Commun 666 (b) Sammes PG, Street LJ, Whitby RJ (1986) J Chem Soc Perkin Trans I 281
58. (a) Wender PA, Lee HY, Wilhelm RS, Williams PD (1989) J Am Chem Soc 111:8954 (b) Garst ME, McBride BJ, Douglass III JG (1983) Tetrahedron Lett 24:1675
59. Williams DR, Benbow JW, Allen EE (1990) Tetrahedron Lett 31:6769
60. Lupi A, Patamia M, Aramone F (1990) Gazz Chim Ital 120:277
61. Wender PA, McDonald FE (1990) J Am Chem Soc 112:4956
62. Wender PA, Mascarenas JL (1991) J Org Chem 56:6267
63. Feldman KS (1983) Tetrahedron Lett 24:5585
64. Padwa A, Weingarten MD (1996) Chem Rev 96:223
65. Ibata T, Jitsuhiro K, Tsubokura Y (1981) Bull Chem Soc Jpn 54:240
66. Padwa A, Carter SP, Nimmesgern H (1986) J Org Chem 51:1157
67. Padwa A, Fryxell GE, Zhi L (1988) J Org Chem 53:2875
68. Padwa A, Fryxell GE, Zhi L (1990) J Am Chem Soc 112:3100
69. Padwa A, Carter SP, Nimmesgern H, Stull PD (1988) J Am Chem Soc 110:2894
70. Padwa A, Hornbuckle SF, Fryxell GE, Stull PD (1989) J Org Chem 54:817
71. McMills MC, Zhuang L, Wright DL, Watt W (1994) Tetrahedron Lett 35:8311
72. Padwa A, Sandanayaka VP, Curtis EA (1994) J Am Chem Soc 116:2667
73. Padwa A, Chinn RL, Hornbuckle SF, Zhang ZJ (1991) J Org Chem 56:3271
74. Pirrung MC, Werner JA (1986) J Am Chem Soc 108:6060
75. West FG, Eberlein TH, Tester RW (1993) J Chem Soc Perkin Trans 1 2857
76. (a) Molander GA, Shubert DC (1987) J Am Chem Soc 109:6877 (b) Molander GA, Andrews SW (1989) Tetrahedron Lett 30:2351
77. (a) Brownbridge P, Chan TH (1979) Tetrahedron Lett 4437 (b) Lee SD, Chan TH (1984) Tetrahedron 40:3611
78. Molander GA, Cameron KO (1991) J Org Chem 56:2617
79. Molander GA, Cameron KO (1993) J Am Chem Soc 115:830
80. Molander GA, Cameron KO (1993) J Org Chem 58:5931
81. Molander GA, Eastwood PR (1995) J Org Chem 60:8382
82. Molander GA, Siedem CS (1995) J Org Chem 60:130
83. Kobayashi K, Sasaki A, Kanno Y, Suginome H (1991) Tetrahedron 47:7245
84. Molander GA, McKie JA (1993) J Org Chem 58:7216
85. Harmata M, Murray T (1989) J Org Chem 54:3761
86. Davies SG, Polywka MEC, Thomas SE (1986) J Chem Soc Perkin Trans I 1277
87. (a) Alvarez E, Diaz MT, Rodriguez ML, Martin JD (1990) Tetrahedron Lett 31:1629 (b) Zarraga M, Martin JD (1991) Tetrahedron Lett 32:2249
88. Alvarez E, Zurita D, Martin JD (1991) Tetrahedron Lett 32:2245
89. Rigby JH, Zbur Wilson JA (1987) J Org Chem 52:34
90. Gebel RC, Margaretha P (1992) Helv Chim Acta 75:1633
91. Fattori D, Vogel P (1993) Tetrahedron Lett 34:1017
92. Schmidt RR, Beitzke C, Forrest AK (1982) J Chem Soc Chem Commun 909
93. Le Drian C, Vieira E, Vogel P (1989) Helv Chim Acta 72:338
94. Le Drian C, Vogel P (1987) Helv Chim Acta 70:1703
95. (a) Auberson Y, Vogel P (1989) Helv Chim Acta 72:278 (b) Nativi C, Reymond JL, Vogel P (1989) Helv Chim Acta 72:882 (c) Warm A, Vogel P (1986) J Org Chem 51:5348
96. Takahashi T, Iyobe A, Arai Y, Koizumi T (1989) Synthesis 189
97. Brown HC, Vara Prasad JVN (1985) J Org Chem 50:3002
98. Rama Rao AV, Yadav JS, Vidyasagar V (1985) J Chem Soc Chem Commun 55
99. Arjona O, Fernandez de la Pradilla R, Perez RA, Plumet J (1988) Tetrahedron 44:7199
100. (a) La Belle BE, Knudsen MJ, Olmstead MM, Hope H, Yanuch MD, Schore NE (1985) J Org Chem 50:5215 (b) Sampath V, Schore NE (1985) J Org Chem 48:4882
101. Fattori D, de Guchteneere E, Vogel P (1989) Tetrahedron Lett 30:7415
102. Black KA, Vogel P (1986) J Org Chem 51:5341

103. (a) Reymond JL, Vogel P (1989) Tetrahedron Lett 30:705 (b) Reymond JL, Pinkerton AA, Vogel P (1991) J Org Chem 56:2128
104. Arjona O, Fernandez de la Pradilla R, Garcia L, Mallo A, Plumet J (1989) J Chem Soc Perkin Trans II 1315
105. (a) Arjona O, Fernandez de la Pradilla F, Garcia E, Martin-Domenech A, Plumet J (1989) Tetrahedron Lett 30:6437 (b) Arjona O, Fernandez de la Pradilla R, Martin-Domenech A, Plumet J (1990) Tetrahedron 46:8187
106. (a) Arjona O, Fernandez de la Pradilla R, Mallo A, Perez S, Plumet J (1989) J Org Chem 54:4158 (b) Arjona O, Fernandez de la Pradilla R, Manzano C, Perez S, Plumet J Tetrahedron Lett (1987) 28:5547
107. Arjona O, Fernandez de la Pradilla R, Perez S, Plumet J (1988) Tetrahedron 44:1235
108. Moursounidis J, Wege D (1983) Aust J Chem 36:2473
109. (a) Gasparini F Vogel P (1989) Helv Chim Acta 72:271 (b) Bialecki M, Vogel P (1994) Tetrahedron Lett 35:5213 (c) Bialecki M, Vogel P (1995) Helv Chim Acta 78:325
110. Ager DJ, East MB (1994) Heterocycles 37:1789
111. (a) Katagiri N, Akatsuka H, Kaneko C, Sera A (1988) Tetrahedron Lett 29:5397 (b) Katagiri N, Akatsuka H, Haneda T, Kaneko C (1987) Chem Lett 2257
112. Lautens M, Abd-El-Aziz AS, Lough AJ (1990) J Org Chem 55:5305
113. Lautens M, Ma S, Yee A (1995) Tetrahedron Lett 36:4185
114. White JD, Fukuyama Y (1979) J Am Chem Soc 101:226
115. Kim H, Ziani-Cherif C, Oh J, Cha JK (1995) J Org Chem 60:792
116. Lautens M, Belter RK (1992) Tetrahedron Lett 33:2617
117. (a) Arjona O, de Dios A, Fernandez de la Pradilla R, Plumet J, Viso A (1994) J Org Chem 59:3906 (b) Sammes PG, Street LJ (1983) J Chem Soc Perkin Trans I 2729
118. Williams DR, Benbow JW, McNutt JG, Allen EE (1995) J Org Chem 60:833
119. Ager DJ, East MB (1993) Tetrahedron 49:5683
120. (a) Lautens M, Ma S (1996) Tetrahedron Lett 37:1727 (b) Uozumi Y, Hayashi T (1993) Tetrahedron Lett 34:2335
121. (a) Bunn BJ, Cox PJ, Simpkins NS (1993) Tetrahedron 49:207 (b) Simpkins NS (1996) Pure & Appl Chem 68:691
122. Sweger RW, Czarnik AW (1991) Retrograde Diels-Alder Reactions in Trost BM, Fleming I (eds) Comprehensive Organic Synthesis. vol 5, Pergamon, Oxford, p 551
123. (a) Lautens M (1993) Synlett 177 (b) Lautens M (1993) Pure & Appl Chem 64:1873 (c) Lautens M, Ren Y, Delanghe PHM, Chiu P, Ma S, Colucci J (1995) Can J Chem 73:1251 (d) Keay BA, Woo S (1996) Synthesis 669
124. (a) Arvai G, Fattori D, Vogel P (1992) Tetrahedron 48:10621 (b) Roser K, Carrupt PA, Vogel P, Honegger E, Heilbronner E (1990) Helv Chim Acta 73:1
125. de Guchteneere E, Fattori D, Vogel P (1992) Tetrahedron 48:10603
126. Noyori R, Sato T, Hyakawa Y (1978) J Am Chem Soc 100:2561
127. Arco MJ, Trammell MH, White JD (1976) J Org Chem 41:2075
128. Hoffmann HMR (1984) Angew Chem Int Ed Engl 23:1
129. Sato T, Hayakawa Y, Noyori R (1984) Bull Chem Soc Jpn 57:2515
130. Auberson Y, Vogel P (1989) Angew Chem Int Ed Engl 28:1498
131. Bimwala RM, Vogel P (1992) J Org Chem 57:2076
132. Sevin AF, Vogel P (1994) J Org Chem 59:5920
133. Cowling AP, Mann J, Usmani AA (1981) J Chem Soc Perkin Trans I 2116
134. Bimwala M, Vogel P (1989) Helv Chim Acta 72:1825
135. Alvarez E, Diaz MT, Perez R, Martin JD (1991) Tetrahedron Lett 32:2241
136. Shizuri Y, Nishiyama S, Shigemori H, Yamamura S (1985) J Chem Soc Chem Commun 292
137. Klein LL (1985) J Am Chem Soc 107:2573
138. Meinwald J (1977) Pure & Appl Chem 49:1275
139. Just G, Liak TJ, Lim M-I, Potvin P, Tsantrizos YS (1980) Can J Chem 58:2024
140. Murai A, Takahashi K, Taketsuru H, Masamune T (1981) J Chem Soc Chem Commun 221
141. (a) Ohno M, Ito Y, Arita F, Shibata T, Adachi K, Sawai H (1984) Tetrahedron 40:145 (b) Shimizu M, Matsukawa K, Fujisawa T (1993) Bull Soc Chem Jpn 66:2128 (c) Matsuki K,

Inoue H, Takeda M (1993) Tetrahedron Lett 34:1167 (d) Jones JB, Francis CJ (1984) Can J Chem 62:2578 e) Bloch R, Gibe-Jampel E, Girard C (1985) Tetrahedron Lett 26:4087 (f) Das J, Hanslanger MF, Gougoutas JZ, Malley MF (1987) Synthesis 1100 (g) Seebach D, Jaeschke G, Wang YM (1995) Angew Chem Int Ed Engl 34:2395

142. Kozikowski AP, Ames A (1981) J Am Chem Soc 103:3923
143. Schlessinger RH, Pettus TRR (1994) J Org Chem 59:3246
144. Katagiri N, Akatsuka H, Haneda T, Kaneko C (1988) J Org Chem 53:5464
145. Padwa A, Zhi L, Fryxell GE (1991) J Org Chem 56:1077
146. (a) Imagawa T, Sugita S, Akiyama T, Kawanisi M (1981) Tetrahedron Lett 22:2569 (b) Akiyama T, Fujii T, Ishiwari H, Imagawa T, Kawanisi M (1978) Tetrahedron Lett 2165
147. (a) Imagawa T, Nurai H, Akiyama T, Kawanisi M (1979) Tetrahedron Lett 1691 (b) Imagawa T, Sonobe T, Ishiwari H, Akiyama T, Kawanisi M (1980) J Org Chem 45:2005
148. (a) Harmata M, Gamlath CB, Barnes CL (1990) Tetrahedron Lett 31:5981 (b) Harmata M, Gamlath CB, Barnes CL (1995) J Org Chem 60:5077
149. Wang WB, Roskamp EJ (1992) Tetrahedron Lett 33:7631
150. Gravel D, Deziel R, Brisse F, Hechler L (1981) Can J Chem 59:2997
151. Garver LC, van Tamelen EE (1982) J Am Chem Soc 104:867
152. Van Royen LA, Mijngheer R, De Clerq PJ (1983) Tetrahedron Lett 24:3145
153. Rajapaksa D, Keay BA, Rodrigo R (1984) Can J Chem 62:826
154. Campbell MM, Kaye AD, Sainsbury M (1983) Tetrahedron Lett 24:4745
155. Campbell MM, Kaye AD, Sainsbury M, Yavarzedeh R (1984) Tetrahedron 40:2461
156. Brion F (1982) Tetrahedron Lett 5299
157. Koreeda M, Jung KY, Ichita J (1989) J Chem Soc Perkin Trans I 2129
158. Leroy J, Fischer N, Wakselman C (1990) J Chem Soc Perkin Trans I 1281
159. Grootaert WM, De Clerq PJ (1986) Tetrahedron Lett 27:1731
160. Takahashi T, Kotsubo H, Iyobe A, Namiki T, Koizumi T (1990) J Chem Soc Perkin Trans I 3065
161. Kinder Jr FR, Bair KW (1994) J Org Chem 59:6965
162. Yang W, Koreeda M (1992) J Org Chem 57:3836
163. Guildford A, Turner RW (1983) J Chem Soc Chem Commun 466
164. Acena JL, Arjona O, Fernandez de la Pradilla R, Plumet J, Viso A (1992) J Org Chem 57:1945
165. (a) Metz P, Cramer E (1993) Tetrahedron Lett 34:6371 (b) Metz P, Stölting J, Läge M, Krebs B (1994) Angew Chem Int Ed Engl 33:2195
166. Molander GA, Eastwood PR (1995) J Org Chem 60:4559
167. Arjona O, de Dios A, Fernandez de la Pradilla R, Plumet J (1991) Tetrahedron Lett 32:7309
168. Le Drian C, Vionnet JP, Vogel P (1990) Helv Chim Acta 73:161
169. Arjona O, Conde S, Plumet J, Viso A (1995) Tetrahedron Lett 34:6157
170. Bhatt MV, Kulkarni SU (1983) Synthesis 249
171. (a) Eggelte TA, de Koning H, Huisman HO (1979) Rec Trav Chim Pays-Bas 98:267 (b) Antonsson T, Vogel P (1990) Tetrahedron Lett 31:89
172. Wong HNC, Ng TK, Wong TY, Xing YD (1984) Heterocycles 22:875
173. (a) Lajunen M, Kaitaranta E, Dahlqvist M (1994) Acta Chem Scand 48:399 (b) Lajunen M, Uotila R (1992) Acta Chem Scand 46:968 (c) Lajunen M, Maki E (1991) Acta Chem Scand 45:578
174. Yates PY, Douglas SP (1982) Can J Chem 60:2760
175. Ogawa S, Iwasawa Y, Taisuke N, Suami T, Ohba S, Ito M, Saito Y (1985) J Chem Soc Perkin Trans I 903
176. Ogawa S, Tsunoda H (1992) Liebigs Ann Chem 637
177. (a) Smith AB, Liverton NJ, Hrib, NJ, Sivaramakrishnan H, Winzenberg K (1985) J Org Chem 50:3239 (b) Smith AB, Liverton NJ, Hrib NJ, Sivaramakrishnan H, Winzenberg K (1986) J Am Chem Soc 108:3040
178. Best WM, Wege D (1981) Tetrahedron Lett 22:4877

179. Giles RGF, Hughes AB, Sargent MV (1991) J Chem Soc Perkin Trans 1 1581
180. Batt DG, Jones DG, La Greca S (1991) J Org Chem 56:6704
181. Takaya H, Hayakawa Y, Makino S, Noyori R (1978) J Am Chem Soc 100:1778
182. (a) Kato T, Suzuki T, Ototani N, Maeda H, Yamada K, Kitahara Y (1977) J Chem Soc Perkin Trans I 206 (b) Kitahara Y, Kato T, Ototani N, Inoue A, Izumi H (1968) J Chem Soc (C) 2508
183. Borthwick AD, Curry DJ, Poynton A, Whalley WB, Hooper JW (1980) J Chem Soc Perkin Trans I 2435
184. Koreeda M, Gopalaswamy R (1995) J Am Chem Soc 117:10595
186. (a) Montana AM, Nicholas KM, Khan MA (1988) J Org Chem 53:5193 (b) Montana AM, Nicholas KM (1990) J Org Chem 55:1569
187. (a) Cummins WJ, Drew MGB, Mann J, Markson AJ (1988) Tetrahedron 44:5151 (b) de Almeida Barbosa L-C, Mann J (1990) J Chem Soc Perkin Trans I 177
188. Dienes Z, Antonsson T, Vogel P (1993) Tetrahedron Lett 34:1013
189. Barbosa LCA, Mann J, Wilde PD (1989) Tetrahedron 45:4619
190. Stohrer I, Hoffmann HMR (1992) Tetrahedron 48:6021
191. (A) Ashworth RW, Berchtold GA (1977) Tetrahedron Lett 339 (b) Hogeveen H, Middelkoop TB (1973) Tetrahedron Lett 3671
192. (a) Grieco PA, Zelle RE, Lis R, Finn J (1983) J Am Chem Soc 105:1403 (b) Grieco PA, Lis R, Zelle RE, Finn J (1986) J Am Chem Soc 108:5908
193. Forsey SP, Rajapaksa D, Taylor NJ, Rodrigo R (1989) J Org Chem 54:4280
194. Pelter A, Ward RS, Li Q, Pis J (1994) Tetrahedron Asymm 5:909
195. (a) De Geyter T, Cauwberghs S, De Clercq (1994) Bull Soc Chim Belg 103:433 (b) Cauwberghs SG, De Clercq PJ (1988) Tetrahedron Lett 29:6501
196. De Schrijver J, De Clerq PJ (1993) Tetrahedron Lett 34:4369
197. Cossy J, Ranaivosata JL, Bellosta V, Ancerewicz J, Ferritto R, Vogel P (1995) J Org Chem 60:8351
198. Jung ME, Street LJ (1984) J Am Chem Soc 106:8327
199. Wender PA, Kogen H, Lee HY, Munger Jr JD, Wilhelm RS, Williams PD (1989) J Am Chem Soc 111:8957
200. (a) Yadav JS, Ravishankar R, Lakshman S (1994) Tetrahedron Lett 35:3617 (b) Yadav JS, Ravishankar R, Lakshman S (1994) Tetrahedron Lett 35:3621
201. Lautens M, Chiu P (1991) Tetrahedron Lett 32:4827
202. Lautens M, Smith AC, Abd-El-Aziz A, Huboux AH (1990) Tetrahedron Lett 31:3253
203. Krishnamurthy S, Brown HC (1979) J Org Chem 44:3678
204. Moss RJ, Rickborn B (1985) J Org Chem 50:1381
205. Lautens M, Chiu P, Colucci JT (1993) Angew Chem Int Ed Engl 32:281
206. Woo S, Keay BA (1992) Tetrahedron Lett 33:2661
207. (a) Lautens M, Chiu P, Ma S, Rovis T (1995) J Am Chem Soc 117:532 (b) The reaction conditions have been optimized (< 2mol % catalyst), T. Rovis, U. Toronto, submitted for publication
208. Lautens M, Ma S (1997) J Am Chem Soc 119:0000
209. (a) Lautens M, Klute W (1996) Angew Chem Int Ed Engl 35:442 (b) Lautens M, Kumanovic S, Meyer C (1996) Angew Chem Int Ed Engl 35:1329
210. Cossy J, Aclinou P, Bellosta V, Furet N, Baranne-Lafont J, Sparfel D, Souchaud C (1991) Tetrahedron Lett 32:1315
211. Caple R, Chen GMS, Nelson JD (1971) J Org Chem 36:2874
212. Jeffrey AM, Yeh HJC, Jerina DM, DeMarinis RM, Foster CH, Piccolo DE, Berchtold GA (1974) J Am Chem Soc 96:6929
213. Arjona O, de Dios A, Plumet J (1993) Tetrahedron Lett 34:7451
214. Arjona O, Fernandez de la Pradilla R, Mallo A, Plumet J, Viso A (1990) Tetrahedron Lett 31:1475
215. Acena JL, Arjona O, Iradier F, Plumet J (1996) Tetrahedron Lett 37:105
216. (a) Lautens M, Chiu P (1993) Tetrahedron Lett 34:773. (b) Lautens M, Fillion E (1996) J Org Chem 61:7994 and references to earlier examples of the "pincer" Diels-Alder reaction

217. (a) Metz P, Meiners U, Fröhlich R, Grehl M (1994) J Org Chem 59:3687 (b) Metz P, Meiners U, Cramer E, Fröhlich R, Wibbeling B (1996) Chem Commun 431
218. Lautens M, Ma S, Belter RK, Chiu P, Leschziner A (1992) J Org Chem 57:4065
219. Lautens M, Gajda C (1993) Tetrahedron Lett 34:4591
220. Lautens M, Kumanovic S (1995) J Am Chem Soc 117:1954
221. Lautens M, Gajda C, Chiu P (1993) J Chem Soc Chem Commun 1193
222. Lautens M, Di Felice C, Huboux A (1989) Tetrahedron Lett 30:6817
223. Lautens M, Belter RK, Lough AJ (1992) J Org Chem 57:422
224. Duan JP, Cheng CH (1993) Tetrahedron Lett 34:4019
225. Moinet C, Fiaud JC (1995) Tetrahedron Lett 36:2051
226. Lautens M, Ma S (1996) J Org Chem 61:7246

The Nucleophilic Addition/Ring Closure (NARC) Sequence for the Stereocontrolled Synthesis of Heterocycles

Patrick Perlmutter

Department of Chemistry, Monash University, Melbourne, Victoria, 3168 Australia

This review brings together examples from the recent literature which demonstrate the potential of nucleophilic addition/ring closure (NARC) sequences for the synthesis of heterocyclic compounds. A heavy emphasis is placed on the stereoselectivity associated with such syntheses. After an introductory section the material is organised into a series of sections based on different classes of nucleophiles. The first (and major) section deals with nucleophilic additions to aldehydes, ketones and aldimines. High levels of stereocontrol in both the nucleophilic addition step (especially where the nucleophile is a chiral enolate) and the ring closure step (which often involves electrophilic activation) are often obtained. Examples are given in the areas of naturally-occurring tetrahydrofurans and tetrahydropyrans. In the final section examples of NARC sequences involving lactones are given.

Table of Contents

1
Introduction

This review brings together examples from the recent literature which demonstrate the potential of nucleophilic addition/ring closure (NARC) sequences for the stereocontrolled synthesis of heterocyclic compounds [1]. This Chapter will largely restrict itself to ring closures onto alkenes. This allows, for the most part,

Topics in Current Chemistry, Vol. 190
© Springer Verlag Berlin Heidelberg 1997

the direct introduction of a second (and, sometimes, a third) new stereogenic centre. (Conceptually, there is no reason why similar processes involving alkynes cannot be developed as the resulting products can be
converted into new stereocentres in subsequent reactions, e.g. diastereoselective reduction). The potential in this approach lies mainly in the combination of any one of a large variety of stereoselective nucleophilic addition
processes with one of an increasingly large number of methods of ring
closure. To date only a very small number of these combinations has been
reported.

The NARC process is represented, schematically, in Fig. 1. Two basic approaches may be taken. In the first, the nucleophile is added to a carbonyl or carbonyl

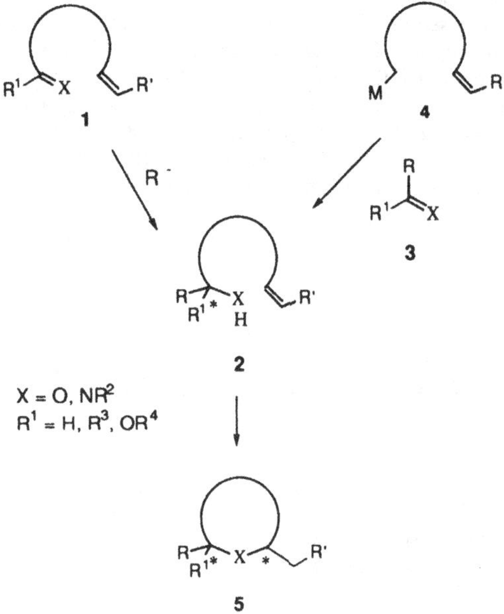

Fig. 1. General scheme showing the two main approaches used in the
nucleophilic addition/ring closure (NARC) process

derivative which is attached to a remote double bond (e.g. 1 → 2). In the second,
the nucleophile which contains a remote double bond (e.g. 4) is added to a carbonyl or carbonyl derivative (e.g. 3 → 2). The product is, in principle, the same
(i.e. 2) and can then be closed using a variety of methods. The stereoselectivity
of each of these processes may be controlled by chiral non-racemic auxiliaries,
chiral non-racemic catalysts or chiral non-racemic substrates. Some examples of
these are given in the following sections.

2
Additions to Aldehydes, Ketones and Aldimines

2.1
Amide Enolates

The development of the asymmetric aldol reaction [2] has been dominated by the stereo-controlled addition of chiral, amide-derived enolates to, mainly, aldehydes. This constitutes an excellent method for the first step of many NARC processes. The pamamycins [3] and the nactins [4] are two groups of naturally-occurring ionophores. They contain tetrahydrofuran sub-units which have proved to be suitable targets for the application of the NARC process.

The pamamycins are macrodiolides possessing three *cis*-fused 2,5-disubstituted tetrahydrofurans, two of which form part of a sixteen-membered macrocycle. Our efforts so far have focused on the C1′–C11′ sub-unit of paramycin 607. In this analysis the nucleophilic addition process is an aldol reaction [5] and the ring closure obviously requires alkene activation by an electrophile of some kind. Based on our studies of simpler systems [6] it is now recognised that, in order to introduce the correct stereochemistry at C6′ of **10** (pamamycin numbering), the stereochemistry at C8′ of **6** needed to be (R). Although the stereochemistry of the natural product is (S) at C8′ this was not seen as a problem as (i) the C8′-epimer may serve as the synthetic intermediate for coupling to the other sub-unit (C1–C18) of pamamycin 607 and (ii) if required, inversion of the stereochemistry at C8′ is straightforward. The synthesis of **10** is shown in Fig. 2.

(i) Y*N ... (**7**), CH_2Cl_2, -78°C, 54%; (ii) (a) $Hg(OAc)_2$, CH_3CN, rt (b) Aq. NaCl, 85%;
(iii) AIBN, Bu_3SnH, toluene, 93%; (iv) (a) LiOH, H_2O_2 (b) CH_2N_2 46%; (v) TBAF, THF, 56%

Fig. 2. The synthesis of a C1′–C11′ synthon of pamamycin 607

As the reasons for the diastereoselectivity of *syn*-selective aldol reactions are well established we will focus on the selectivity of the ring closure. We have carried out studies on intramolecular oxymercurations of a series of simple alkenols related to **8** and have found that they consistently close to give predominantly *syn* and not *anti* products (Fig. 3).

This selectivity was accounted for by assuming that the predominant reactive conformation is **A** where the allylic hydrogen of the stereocentre is eclipsing (or close to eclipsing) the alkene (Fig. 4). Complexation by the incoming mercuronium is then hindered by the allylic alkyl group and so complexation occurs from the opposite face (**D**). Subsequent ring closure of **D** then gives the preferred *syn*-diastereomer. A similar mechanism is presumably operating in the ring closure of **8**.

Walkup's group has published a series of papers describing the synthesis of pamamycin and nactin sub-units [7]. A key reaction in their NARC sequence involves a stereoselective ring closure onto an *allene*. As is apparent from Fig. 5 this approach constructs the ring from the opposite end to that shown in Fig. 2. The high *cis*-selectivity in the ring closure is apparently controlled by the silyl ether moiety. An example of their chemistry is outlined in Fig. 5.

Fig. 3. Diastereoselective intramolecular oxymercurations of alkenols bearing a remote allylic ether

Fig. 4. Likely reactive conformations of alkenols bearing a remote allylic ether

(i) X* $\overset{OBBu_2}{\underset{N}{\diagup\!\diagdown}}$ (**12**), CH$_2$Cl$_2$, -78°C, 75%; (ii) TMSCl, Et$_3$N, 92%; (iii) (a) Hg(OCOCF$_3$)$_2$, CH$_2$Cl$_2$, 25°C (b) CO, MeOH, PdCl$_2$, CuCl$_2$, CH$_3$C(OEt)$_3$, propylene oxide, 85%; (iv) LiOH, H$_2$O$_2$, 90%; (v) BH$_3$THF, 80%; (vi) Mg, MeOH, 58%; (vii) PCC, Ch$_2$Cl$_2$, 90%; (viii) allyltrimethylsilane TiCl$_4$, CH$_2$Cl$_2$, 84%; (ix) H$_2$, Pd-C, E + OH, 71%

Fig. 5. The synthesis of the C1′–C11′ synthon of pamamycins 635 and 649 B

In principle, our approach to the synthesis of pamamycin sub-units should also work well for the preparation of sub-units of nonactin [8]. However, their synthesis requires an *anti*-aldol for the nucleophilic addition step. Until very recently this proved impossible to achieve as aldehyde **16** decomposed in the presence of the strong Lewis acids normally required for this process [9]. We have now established that both the *syn* and the *anti*-aldol products may be obtained with **16** simply by controlling the amount of diethylboron triflate present in the reaction. Thus, addition of boron enolate **17** to **16** gives the expected *syn*-aldol product **18** (Fig. 6) which can then be processed through to diastereomers of nonactate [10]. However addition of an excess of diethylboron triflate, the Lewis acid used in the preparation of the enolate, leads to a new, tandem in-situ NARC process producing **19** in good yield and good diastereoselectivity [11]. This process is all the more remarkable in that the first step is a completely *anti*-selective aldol reaction. The mechanism of this reaction is currently under investigation.

Evans' group has reported the total synthesis of X-206 [12]. A critical aspect of their synthesis was construction of the 2,3,6-trisubstituted tetrahydropyran ring (ring A) using the sequence of (i) aldol followed by (ii) intramolecular oxymercuration. The aldol reaction in this case has a potential added complication to those described for pamamycin above in that the aldehyde has a stereocentre at C2 (see **20** in Fig. 7). This could lead to "substrate" rather than "reagent" control. However the auxiliary completely dominated the stereoselectivity yielding a single diastereomer in almost quantitative yield

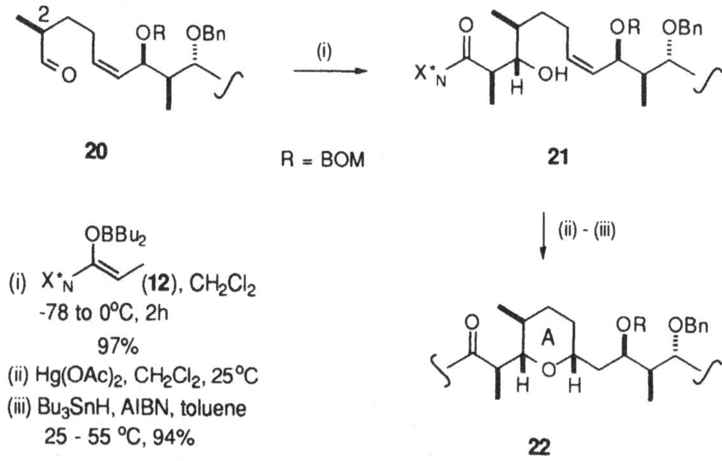

Fig. 6. Synthesis of nonactate precursors using either *syn*- or *anti*-selective aldol reactions

(Fig. 7). The ring closure also proved to be remarkably stereoselective. Thus intramolecular oxymercuration, followed by reductive demercuration, provided the tetrahydropyran (**22**, ring A of X-206) with the desired 2,6-*cis*-relative stereochemistry in excellent overall yield as a single diastereomer.

The authors suggest that the very high diastereoselection in this ring closure is due to a combination of conformational effects. In essence, a transition state (**23**) which involves a chair-like conformation *and* has the hydrogen attached to the remote allylic centre "eclipsing" the double bond, should be the most favourable for ring closure (Fig. 8). This certainly accounts for the diastereoselectivity observed and is supported by a series of model studies [13].

Fig. 7. Evans' synthesis of ring A of X-206

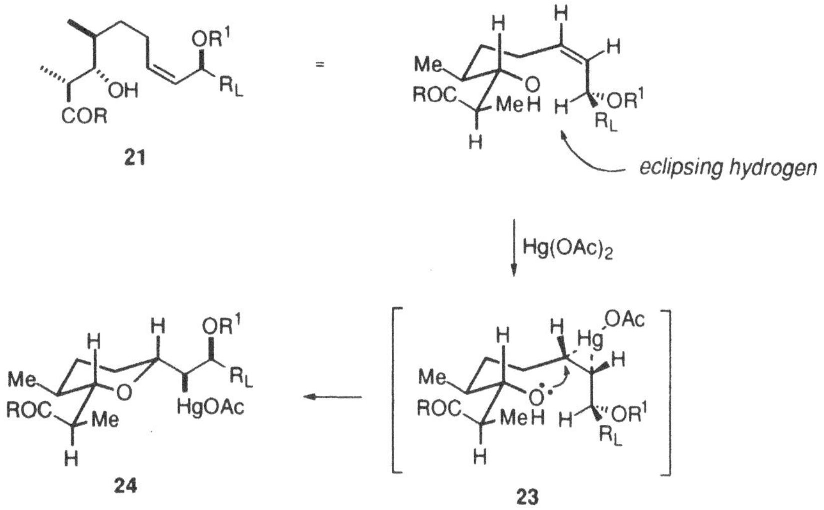

Fig. 8. Evans' mechanism for the diastereoselection observed in the ring closure of **21**

2.2
Ester Enolates

Galatsis' group [14] reported a study on an NARC sequence involving (i) aldol reactions of enolates derived from the kinetic deprotonation of unsaturated esters, such as **25** and **28**, to ketones (Fig. 9) and aldehydes (Fig. 10) followed by (ii) *endo*-cyclisation via intramolecular iodoetherification. As the enolates used in the study were racemic and the aldol reactions stereorandom, it would be interesting to repeat this work using a chiral auxiliary (e.g. a chiral amide). This should ensure high levels of enantio- and diastereo-selectivity.

The authors found that the *endo*-cyclisations were mostly highly diastereo-selective which is consistent with previous reports using other hydroxyalkenes.

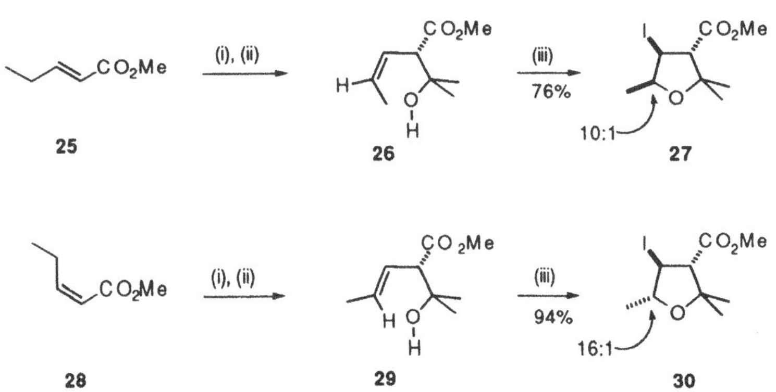

Fig. 9. NARC sequences initiated by aldol additions of ester enolates to acetone

(i) LDA, HMPA; (ii) MeCHO; (iii) I$_2$, NaHCO$_3$, MeCN, rt, 24h

Fig. 10. NARC sequences initiated by aldol additions of ester enolates to acetaldehyde

They rationalise the 3,4-*trans*-selectivity in all cases by assuming that transition structures **39** and **41** are lower in energy than **40** and **42** respectively (Fig. 11).

2.3
Ketone Enolates

The antibiotic calcimycin (or A23187) is a widely used probe for calcium ion transport in biological systems. A synthesis of the core of this antibiotic has been developed [15]. Although little stereoselectivity is associated with this method it is rather remarkable in that *two* ring closures are involved, the first involving hemi-acetal formation and the second an electrophilic closure (Fig. 12).

2.4
Organozinc Reagents

Knochel has developed an effective [3+2] cycloaddition strategy which involves a nucleophilic addition of a *tert*-butylsulfonyl-containing allylzinc reagent with

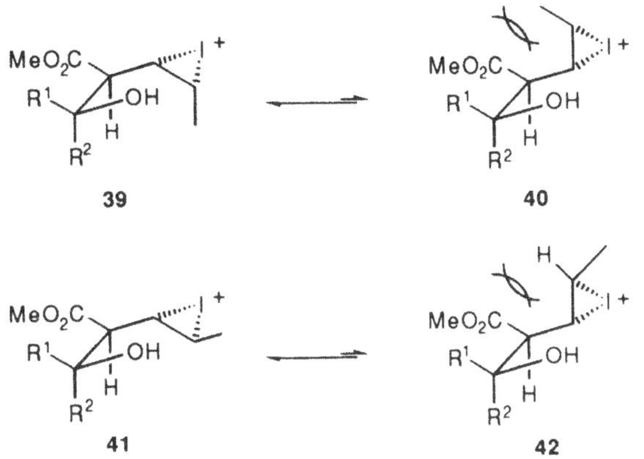
MeO2C
R1
OH
H
R2
39
40
41
42

Fig. 11. Galatsis' transition state analysis of *endo* cyclisations

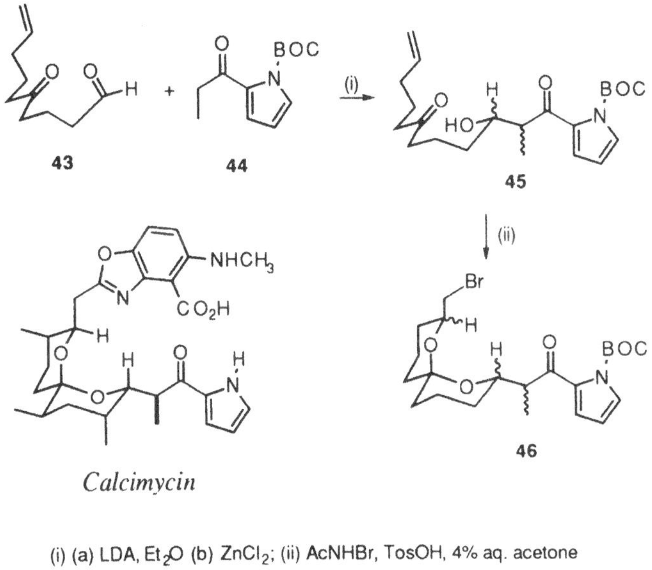
43
44
45
46
BOC
(i)
(ii)
Br
NHCH3
CO2H
Calcimycin
(i) (a) LDA, Et2O (b) ZnCl2; (ii) AcNHBr, TosOH, 4% aq. acetone

Fig. 12. Model calcimycin synthesis

aldehydes or imines (Fig. 13) [16]. The procedure requires the process to be carried out in a stepwise manner as the intermediate zinc alkoxide is apparently not nucleophilic enough to add to the alkene. However, the ring closure is catalysed by potassium hydride in good yield.

The ring closure (**54** → **56**), which is formally a disfavoured 5-endo-trig [17], is all the more remarkable as it can successfully compete against another very fast process, namely an anionic oxy-Cope rearrangement (i.e. **54** → **55**, Fig. 14).

(i) (48), Zn, THF
(ii) KH (cat.), THF, 25 °C, 10 min
88%

(i) (48), Zn, THF, 65%
(ii) KH (cat.), THF, 25 °C, 10 min, 65%

Fig. 13. Knochel's [2+3] cycloaddition process

(i) KH (1 equiv.), THF, 25 °C, 81%

Fig. 14. Ring closure of the oxy-anion derived from 54

2.5
Organosilanes and Stannane Reagents

Trost's group has developed an annulation reagent (a trimethylenemethane syn-thon) which achieves a nucleophilic addition to an aldehyde followed by a ring closure onto a π-allyl palladium complex in the one pot (Fig. 15) [18].

Under the first set of conditions developed for these reactions the reported diastereoselectivity in additions to chiral aldehydes was only modest (Fig. 16) [18].

However a subsequent study by the same group revealed that, by employing the trialkylstannane equivalent (66) to the reagent initially described and employing a strong Lewis acid at low temperatures instead of a palladium cata-

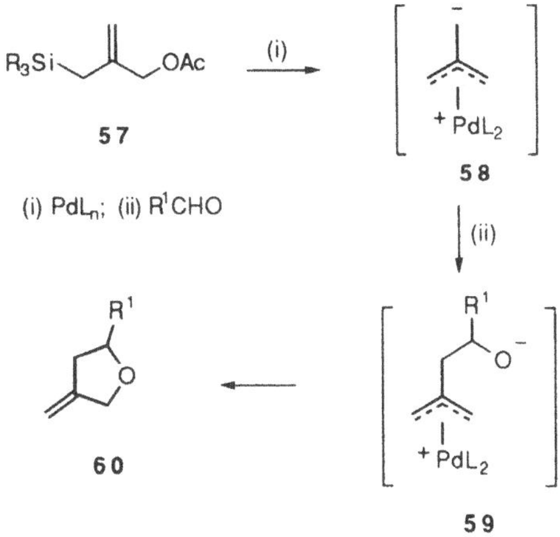

Fig. 15. Trost's design for the annulation of aldehydes with the trimethylenemethane reagent **57**

(i) Pd(OAc)$_2$ (5 mol%), Bu$_3$SnOAc (20 mol%), PPh$_3$ (25 mol%), **57**, THF, reflux

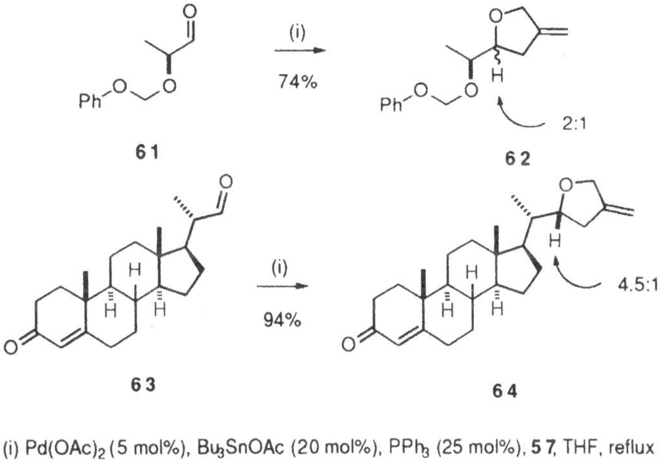

Fig. 16. Trost's one-pot annulations of aldehydes using trimethylenemethane reagent **57**

lyst, good to excellent levels of diastereoselectivity could be achieved in a discrete, nucleophilic addition step (Fig. 17) [19]. Ring closure of the adducts was then completed using palladium catalysis in the presence of a base. The base was necessary as the alcohols were not sufficiently nucleophilic to achieve ring closure.

In the same paper, the authors demonstrated another advantage of this stepwise approach. The initial adduct can be transformed into either diastereomer simply by inverting the reactivity of the two alcohols (Fig. 18).

This process was also applied to the annulation of a series of imines (Fig. 19) [19].

(i) (a) AcO⟶⟶SnBu₃ **(66)**, CH₂Cl₂, BF₃·OEt₂ (3 equiv.), -78°C, 45 min (b) Aq. NH₄Cl;
(ii) (a) Pd(OAc)₂ (0.05 equiv.), PPh₃ (0.25 equiv.), BuLi (0.1 equiv.), dioxane (b) DBU,
reflux

Fig. 17. Diastereoselective annulations using trimethylenemethane reagent **66**

(i) MsCl, Et₃N, CH₂Cl₂, -5°C; (ii) KOH, H₂O/MeOH (5:1), reflux

Fig. 18. Preparation of *epi*-67

(i) See Fig. 17 (ii) Pd(OAc)₂ (0.05 equiv.), PPh₃, THF, n-BuLi (0.1 equiv.)
Et₃N (1.5 equiv.) reflux

Fig. 19. Trost's pyrrolidine synthesis

3
Additions to Lactones

3.1
Organomagnesium Reagents

Tachibana's group has reported its attempts to apply the NARC sequence to the preparation of the spiro-acetal moiety of the ciguatoxins (rings L/M) [20]. This approach relies on the nucleophilic addition to C1 occurring from the β-face (i.e. the face opposite to the C2 methyl group) and a diastereoselective oxidative ring closure. They showed that the nucleophilic addition of allyl-magnesium bromide is completely stereoselective. However ring closure using an osmium(VIII) catalysed dihydroxylation gave a mixture of all four possible diastereomers. Apparently, under the reaction conditions, the hemiacetal (78) equilibrated with the ring open form prior to dihydroxylation (Fig. 20).

77

78

(i) CH$_2$=CHCH$_2$MgBr, THF, -78 °C
   100%
(ii) OsO$_4$, CH$_3$CN, H$_2$O, r.t.
(iii) Na$_2$SO$_3$, 91%

79

Fig. 20. Tachibana's model synthesis of the KLM portion of the ciguatoxins

Shortly after this Tachibana's group published an improved procedure which employs Corey's asymmetric dihydroxylation protocol [21] to install the correct stereochemistry in ring M. Under these conditions there is apparently no hemiacetal ring opening and the intermediate triol closes virtually quantitatively to the spiroacetal. The successful execution of this ring synthesis is given in Fig. 21 [22].

In preliminary studies we have found that this is also a powerful approach to the preparation of enantiomerically pure complex spiroacetals (Fig. 22 [23, 24]). Intramolecular oxymercuration of **84** proceeds efficiently, although without any stereoselectivity.

(i) CH$_2$=CHCH$_2$MgBr, THF, -78°C
79%

(ii) OsO$_4$, TMBHN   NHTMB
                Ph   Ph

(iii) CSA, 97%

TMB = 2,4,6 - trimethylbenzyl

Fig. 21. Tachibana's total synthesis of the KLM portion of the ciguatoxins

(i) CH$_2$=CHCH$_2$CH$_2$MgBr (6 equiv.),
(ii) (a) Hg(OAc)$_2$, CH$_2$Cl$_2$
     (b) Aq. NaCl

Fig. 22. Stereoselective synthesis of spiroacetals

4
The Future

Clearly the NARC sequence is a very powerful one for the preparation of
heterocycles. Although the technology is still in its infancy some significant
applications to the enantioselective synthesis of important target molecules
have already appeared. It seems very likely that many new examples of this
method will appear over the next few years.

References

1. For a previous review, see Perlmutter P (1996) Curr Med Chem 3:139
2. Heathcock H (1984) In: Morrison JD (ed) Asymmetric synthesis. Academic Press, New York vol III part B Chapter 2
3. (a) McCann PA, Pogell BM (1979) J. Antibiotics 32:673; (b) Natsume M, Yasui K, Kondo S, Marumo S (1991) Tetrahedron Lett 32:3087 and references cited therein
4. Corbaz R, Ettlinger L, Gaumann E, Keller-Schierlein, Kradolfer F, Neipp L, Prelog V, Zähner H (1955) Helv Chim Acta 38:1445; (b) Dominquez, J, Dunitz JD, Gerlach H, Prelog V (1962) Helv Chim Acta 45:129; (c) Gerlach H, Prelog V (1963) Liebigs Ann Chem 669:121; (d) Kilbourn BT, Dunitz JD, Pioda LAR, Simon W (1967) J Mol Biol 30:559
5. Evans DA (1982) Aldrichimica Acta 15:23
6. Garavelas A, Mavropoulos I, Perlmutter P, Westman G (1995) Tetrahedron Lett 36:463
7. See Walkup RD, Kim YS (1995) Tetrahedron Lett 36:3091 and references cited therein
8. For example, see Bartlett PA, Meadows JD, Ottow E (1984) J Am Chem Soc 106:5304
9. (a) Raimundo BC, Heathcock, CH (1995) Synlett 1213; (b) Walker MA, Heathcock CH (1991) J Org Chem 56:5747; (c) Danda H, Hansen M, Heathcock CH (1990) J Org Chem 55:173
10. Bratt K, Garavelas A, Perlmutter P, Westman G (1996) J Org Chem 61:2109
11. Hockless DCR, Jones ED, Mavropoulos I, Perlmutter P (1996) J Org Chem 61:submitted
12. Evans DA, Bender SL, Morris J (1988) J Am Chem Soc 110:2506
13. Bender SL (1986) Ph D Thesis, Harvard University
14. Galatsis P, Millan S, Nechala P, Ferguson G (1994) J Org Chem 59:6643
15. Prudhomme M, Dauphin G, Jeminet G (1987) J Chem Res (S) 420
16. Auvray P, Knochel P, Normant JF (1985) Tetrahedron Lett 26:4455
17. Baldwin JE (1976) J Chem Soc, Chem Commun 734 and 738
18. Trost BM, King SA (1986) Tetrahedron Lett 27:5971
19. Trost BM, Bonk PJ (1985) J Am Chem Soc 107:1778
20. Sasaki M, Hasegawa A, Tachibana K (1993) Tetrahedron Lett 34:8489
21. Corey EJ, Jardine DP, Virgil S, Yuen P-W, Connel RD (1989) J Am Chem Soc 111:9243
22. Sasaki M, Masayuki I, Tachibana K (1994) J Org Chem 59:715
23. Guy S, Perlmutter P Unpublished results
24. For a similar sequence involving a free radical ring closure, see Kraus GA, Thurston J (1987) Tetrahedron Lett 28:4011

Chiral Acetylenic Sulfoxides and Related Compounds in Organic Synthesis

Albert W. M. Lee and W. H. Chan

Department of Chemistry, Hong Kong Baptist University, Kowloon Tong, Hong Kong

Sulfoxide, sulfinate and sulfonate are used as activators of acetylenic or vinyl units. Several α, β unsaturated synthons, namely acetylenic sulfoxide (1), vinyl sulfoxide (2), acetylenic sulfinate (3), acetylenic sulfonate (4), and 1-propene-1,3-sultone (5) are developed. Their applications in Diels-Alder reactions, heterocycle and alkaloid syntheses are also investigated. For the chiral acetylenic sulfoxide, the sulfoxide moiety not only enables chemical activation of the acetylene unit, it can also induce stereochemical control at the adjacent carbon centers to achieve enantioselective synthesis.

Table of Contents

Topics in Current Chemistry, Vol. 190
© Springer Verlag Berlin Heidelberg 1997

List of Abbreviations

Ar   aryl
MCPBA   *m*-chloroperoxybenzoic acid
TFA   trifluoroacetic acid
TFAA   trifluoracetic anhydride
TMSOTf   trimethylsilyl trifluoromethanesulfonate
p-Tol   *p*-methylphenyl
Ts   tosyl, *p*-toluenesulfonyl
TsOH   *p*-toluenesulfonic acid

1
Introduction

Sulfoxide, sulfinate and sulfonate are electron-withdrawing groups [1]. They are all capable of stabilizing their corresponding adjacent carbanionic centers. For example, sulfoxide-stabilized α-carbanions have been extensively used for C–C bond formation including asymmetric synthesis [2]. Over the last few years, our research group has been exploring the uses of these sulfur-containing functional groups as activators of acetylenic or vinylic units. Several α, β-unsaturated synthons, namely acetylenic sulfoxide **1**, vinyl sulfoxide **2**, acetylenic sulfinate **3**, acetylenic sulfonate **4**, and propene sultone **5** have been developed and their applications in organic synthesis investigated.

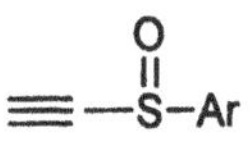

1 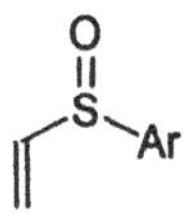   **2** 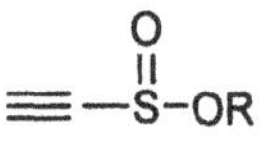   **3** 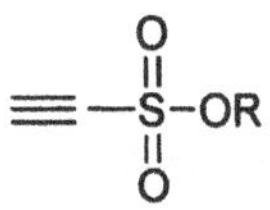   **4** 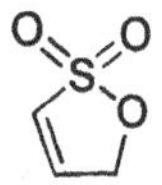   **5**

For the acetylenic sulfoxide, because of its configurationally stable pyramidal stereogenic sulfur atom (a lone electron pair, an oxygen and two different carbon substituents), it can exist in chiral forms. Therefore, in chiral acetylenic sulfoxide, the sulfoxide moiety not only serves as a chemical activator of the acetylene unit, it can also induce stereochemical control at the adjacent carbon centers to achieve enantioselective synthesis. In this article, we shall discuss the preparation of these α, β-unsaturated synthons and their applications in Diels-Alder reactions, heterocycle and alkaloid syntheses.

2
Chiral Acetylenic Sulfoxide

2.1
Synthesis of Homochiral Acetylenic Sulfoxides

There are several efficient methods available for the synthesis of homochiral sulfoxides [3], such as asymmetric oxidation, optical resolution (chemical or biocatalytic) and nucleophilic substitution on chiral sulfinates (the Andersen synthesis). The asymmetric oxidation process, in particular, has received much attention recently. The first practical example of asymmetric oxidation based on a modified Sharpless epoxidation reagent was first reported by Kagan [4] and Modena [5] independently. With further improvement on the oxidant and the chiral ligand, chiral sulfoxides of >95% *ee* can be routinely prepared by these asymmetric oxidation methods. Nonetheless, of these methods, the Andersen synthesis [6] is still one of the most widely used and reliable synthetic route to homochiral sulfoxides. Clean inversion takes place at the stereogenic sulfur center of the sulfinate in the Andersen synthesis. Therefore, the key advantage of the Andersen approach is that the absolute configuration of the resulting sulfoxide is well defined provided the absolute stereochemistry of the sulfinate is known.

Our synthesis of homochiral acetylenic sulfoxides is outlined in Scheme 1. The Grignard reagent of trimethylsilyl acetylene was reacted with sulfinates **6a–c** in toluene. After potassium fluoride desilylation, optically pure acetylenic sulfoxides (*R*)-(+)-**1** were obtained in good yield (Scheme 1) [7,8].

Scheme 1

Since we wanted to prepare a series of chiral acetylenic sulfoxides with different substituents on the aryl moiety, we needed access to the corresponding chiral sulfinates. Optically pure (–)-menthyl-*p*-toluenesulfinate (**6a**) is commercially available but the other sulfinates (**6b** and **6c**) are not. They were prepared according to an efficient procedure developed by Sharpless [9] from substituted benzenesulfonyl chlorides which are commercially available (Scheme 2). The sulfinates were formed as a mixture of diastereomers by *in situ* reduction of the

Scheme 2

sulfonyl chlorides with trimethylphosphite in the presence of triethyl amine and menthol. The diastereomeric sulfinates could be easily separated into optically pure forms by column chromatography or recrystallization from acetone (Scheme 2).

2.2
Enantioselective Alkaloid Synthesis

The application of chiral sulfoxides to the asymmetric synthesis of biologically active compounds has recently been reviewed [3]. Conjugate addition to chiral vinyl sulfoxides has been used by several research groups to achieve the enantioselective synthesis of natural products. For example, intramolecular asymmetric conjugate addition of a nitrogen nucleophile to a chiral vinyl sulfoxide (Scheme 3) was studied by Pyne and applied to the enantioselective synthesis of (R)-carnegine and other alkaloid systems (Scheme 3) [10].

Scheme 3

We view acetylenic sulfoxide **1** as a two-carbon synthon in alkaloid synthesis. Our general approach, as depicted in Scheme 4, called for a Michael addition of Nu^1 to the terminal acetylenic position followed by a cyclization by Nu^2 (an intramolecular second Michael addition). This Michael addition cyclization step will build up the basic skeleton of the alkaloid system and at the same time control the absolute stereochemistry of the newly created chiral center through asymmetric induction of the chiral sulfoxide moiety. Finally, the sulfoxide can be transformed to another functional group (X) or used to promote the formation of another bond with Nu^3 via trapping of the sulfenium ion intermediate under Pummerer rearrangement conditions (Scheme 4).

Scheme 4

2.2.1
Tetrahydroisoquinoline Alkaloids

Our first attempt was an enantioselective synthesis of (R)-$(+)$-carnegine [7, 8]. Michael addition of 2-(3,4-dimethoxyphenyl)ethylamine (**7**) to (R)-$(+)$-**1** took place readily at room temperature in chloroform (Scheme 5). Without isolation of any intermediate, the reaction mixture was treated with excess trifluoroacetic acid to effect the cyclization. Depending on the reaction conditions and the aryl substituent of the chiral sulfoxide, different levels of diastereoselectivity were observed. The results are summarized in Table 1. Under proper conditions (TFA, 0 °C, 4h), **10b** could be obtained in 65% yield as the only isolated product (Scheme 5) (Table 1).

Under the influence of excess TFA, we believed that the Michael addition product **8** should be transformed to the protonated imine form **9** in which hydrogen bonding may exist between the ammonium hydrogen and the sulfoxide oxygen forming a six-membered ring intermediate. We speculate that this intramolecular hydrogen bonding, which locked the conformation of the system, may be res-

Scheme 5

Table 1. Michael addition-cyclization of 7 with chiral acetylenic sulfoxides

Acetylenic Sulfoxide	Acid	T (°C)	Isolated Products (%)	Isolated Yield
1a (Ar = p-Tol)	TFA	0	**10a + 11a** (2 : 1)	45
1b (Ar = o-$NO_2C_6H_4$)	TFA	r.t.	**10b** only	35
1b	TFA	0	**10b** only	65
1b	$BF_2 \cdot Et_2O$	0	**10b** only	20

ponsible for the diastereoselectivity of the cyclization. This speculation was further supported by the fact that acetylenic sulfoxide **1b** afforded a better diastereoselectivity than **1a**, possibly because the electron-withdrawing ortho nitro group in the aryl moiety further stabilized the proposed hydrogen bonding. Boron trifluoride etherate also induced cyclization but the reaction yield was much lower. With reference to our general approach (Scheme 4), the primary amine 7 is Nu[1] for the first Michael addition, and the electron-rich dimethoxyl aryl ring is Nu[2] for the Friedel-Crafts-type cyclization.

We also observed that the reaction time played a crucial part in this reaction sequence. We have evidence that any **11b** formed was actually decomposed in the reaction mixture or during silica gel column chromatography purification. If we ran the reaction with TFA at 0 °C for 4h, **10b** was isolated as the only product in 65% yield. However, if we quenched the reaction mixture with TsCl before it ran to completion at about 2h, some tosylated product of **11** (R = Ts) could be identified.

Reductive amination followed by Raney Nickel desulfurization of **10b** afforded (*R*)-(+)-carnegine (**12**) in good yield.

We further explored the steric effect of this Michael addition-cyclization reaction sequence. A series of secondary amines **13a-f** were prepared and subjected to the Michael addition and acid-induced cyclization (Scheme 6) [12]. The results are summarized in Table 2. In general, we found that the secondary amines were less reactive in this Michael addition-cyclization reaction sequence. The *p*-toluene acetylenic sulfoxide **1a** was not reactive enough and only the stronger electron-withdrawing *o*-nitrophenyl acetylenic sulfoxide **1b** achieved the transformation. In contrast to the primary amine approach, the secondary amine approach resulted in a *reversed* diastereoselectivity bias with compounds **14** as the major isolated products (except **13e**). In general, a lower reaction temperature and increase in the steric hindrance of the secondary amine improved the diastereoselectivity. Exceptionally good diastereoselectivity was observed for the cyclization of **13f** (Scheme 6) (Table 2)

Since reversed diastereoselectivity resulted, starting from the secondary amine **13a**, a convergent synthesis of (*S*)-(−)-carnegine (*ent*-**12**) was achieved by desulfurization of **14a**.

Scheme 6

Table 2. Secondary amine cyclization

Amine	T (°C)	Diastereoselectivity 14:15	Yield (%)
13a	0	1.8 : 1	88
13b	25	2.7 : 1	85
13b	-15	6 : 1	64
13c	0	4.7 : 1	82
13d	0	5.4 : 1	72
13e	0	1 : 4.3	87
13f	40	**14f** only	68

2.2.2
β-Carboline and Yohimbine Alkaloids

Using tryptamine as the nucleophile, the Michael addition-cyclization strategy was extended to the enantioselective synthesis of the β-carboline alkaloid system. Michael addition of tryptamine to the chiral acetylenic sulfoxides took place smoothly at room temperature. Either trifluoroacetic acid or *p*-toluene-sulfonic acid was effective as a catalyst for the cyclization step (Scheme 7). The results of the Michael addition-cyclization reaction sequence are summarized in Table 3. In general, we found that the indole moiety is more reactive than the dimethoxyaryl ring used in the tetrahydroisoquinoline synthesis. Therefore, the cyclization step could take place at a temperature as low as –60 °C. Also, *p*-toluenesulfonic acid resulted in a better diastereoselectivity. However, the diastereoselectivity of the system is much less sensitive to the aryl substituents of the acetylenic sulfoxides compared to that of the tetrahydroisoquinoline system. Also, to our surprise, the steric factor on the chiral acetylenic sulfoxide has little effect on the diastereoselectivity. Even with the bulky 2-methoxy-naphthyl acetylenic sulfoxide **1c** [11], the diastereoselectivity still remained roughly the same as for **1a** and **1b** (Scheme 7) (Table 3).

Diastereomers **17** and **18** could be readily separated by silica gel column chromatography. Raney nickel desulfurization of **17b** completed an enantioselective synthesis of (*R*)-(+)-tetrahydroharman (Scheme 7) [8].

Yohimbine alkaloids possess a characteristic pentacyclic indole skeleton. Representative members of the family include the rauwolfia (reserpine and deserpidine) and the yohimbines. A wide range of medicinal properties has been associated with these compounds and extensive studies have been carried out on the synthesis of the yohimbine alkaloids, including enantioselective syntheses [13, 14]. In our approach, we view the acetylenic sulfoxide as a two-carbon synthon for the C3-C14 segment of the pentacyclic system (see **27**). The chirali-

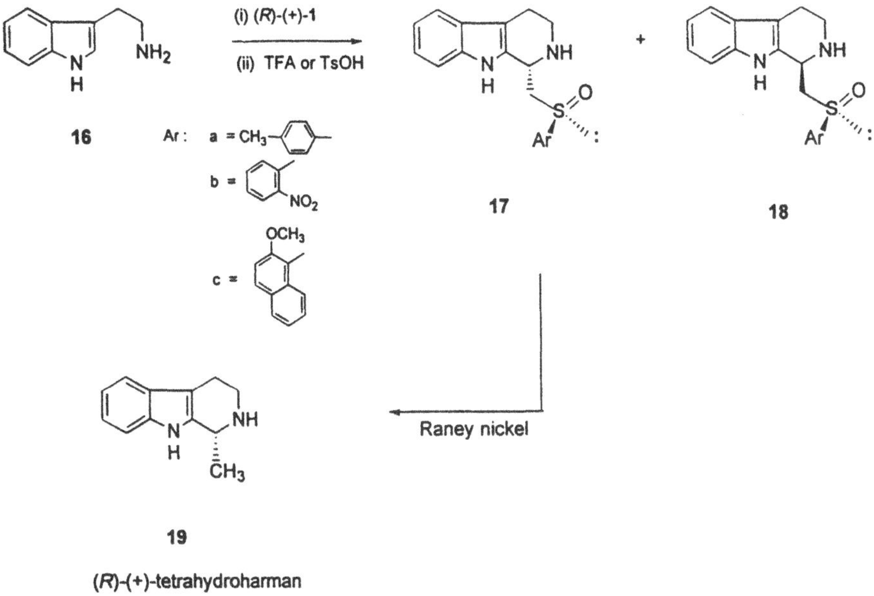

Scheme 7

Table 3. Michael addition cyclization of tryptamine with chiral acetylenic sulfoxides

Sulfoxide	Solvent	Acid	T (°C)	Diastereo-meric ratio 17 : 18	Yield (%)
1a	CHCl$_3$	TFA	-60	3 : 2	85
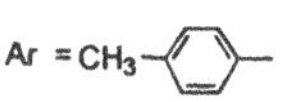 Ar = CH$_3$—	CH$_3$OH	TsOH	-30	7 : 3	91
1b	CHCl$_3$	TFA	-60	7 : 3	60
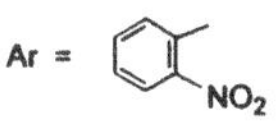 Ar = (NO$_2$)	CH$_3$CN	TFA	-40	7 : 3	60
	CH$_3$OH	TsOH	-30	4 : 1	93
1c	CH$_3$OH	TsOH	0	2 : 1	85
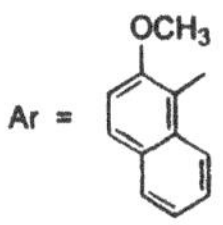 Ar = (OCH$_3$)	CH$_3$OH	TsOH	-30	3 : 1	90
	CH$_3$OH	TsOH	-45	4 : 1	91

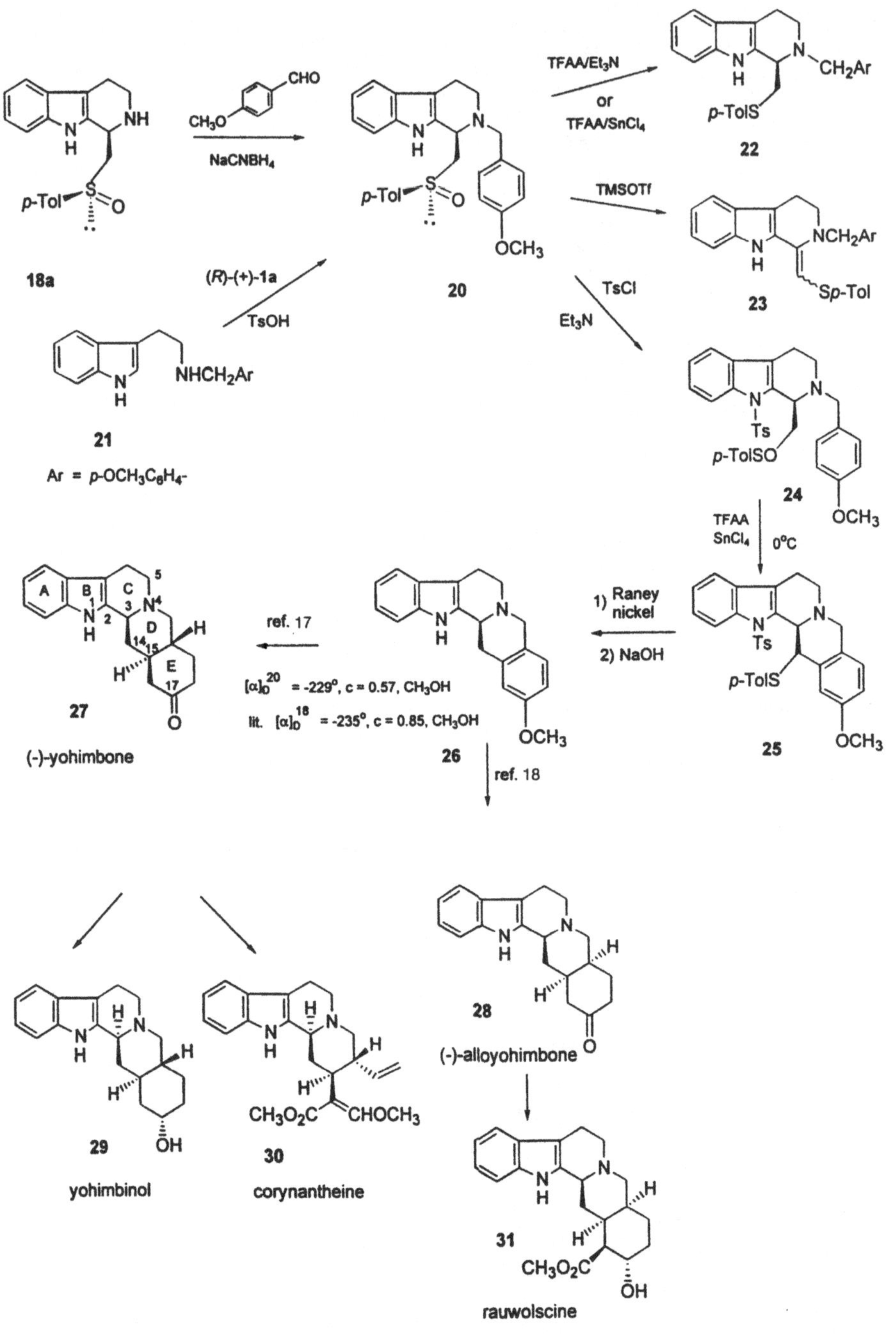
CHO
CH3O
NaCNBH4
18a
(R)-(+)-1a
TsOH
21
Ar = p-OCH3C6H4-
NHCH2Ar
20
OCH3
TFAA/Et3N
or
TFAA/SnCl4
22
p-TolS
CH2Ar
TMSOTf
23
NCH2Ar
Sp-Tol
TsCl
Et3N
24
p-TolSO
Ts
TFAA
SnCl4
0°C
25
p-TolS
1) Raney nickel
2) NaOH
26
ref. 17
[α]D20 = -229°, c = 0.57, CH3OH
lit. [α]D18 = -235°, c = 0.85, CH3OH
27
(-)-yohimbone
ref. 18
28
(-)-alloyohimbone
29
yohimbinol
30
corynantheine
CH3O2C
CHOCH3
31
rauwolscine

Scheme 8

ty of the sulfoxide controls the absolute stereochemistry of the crucial C3 chiral center (Scheme 8).

Compound **20** could be prepared by reductive amination of **18a** with *p*-methoxybenzaldehyde. Alternatively, a secondary amine cyclization approach between **21** and chiral acetylenic sulfoxide **1a** could be used. Again, reversed diastereoselectivity (7:3, 83% yield) in favor of **20**, compared to the primary amine cyclization, was observed. Using the sulfoxide in compound **20** as a handle to effectthe formation of a C14–C15 bond under Pummerer rearrangement conditions proved not to be as straightforward as first anticipated. Upon treatment of compound **20** with typical Pummerer rearrangement reagents, such as trifluoroacetic anhydride or trimethylsilyl triflate [15], only the deoxygenated (**22**) or the elimination (**23**) product could be isolated. Finally, we found that protection of the indole nitrogen is crucial to this transformation. The *N*-tosylated compound **24** underwent Pummerer cyclization in 75% yield with trifluoroacetic anhydride in the presence of tin tetrachloride at 0 °C. This completed the ring D construction (**25**). A new chiral center was also created at C14. Although we have no information about the absolute configuration at C14, only one diastereomer resulted in this cyclization step. Raney nickel desulfurization followed by alkaline detosylation afforded pentacyclic intermediate **26** [16]. Compound **26** was converted to either (–)-yohimbone (**27**) or (–)-alloyohimbone (**28**) through Birch reduction followed by different hydrogenation conditions [17, 18]. Yohimbone was used as a precursor of naturally occurring yohimbol (**29**) and corynantheine (**30**) [17], while rauwolscine (**31**) was synthesized from **28** [18].

Chiral acetylenic sulfoxide **1a** was used as a two-carbon synthon for C3–C14 in building up the pentacyclic alkaloid ring system. Referring back to our general strategy as outlined in Scheme 4, the *p*-methoxy aryl ring served as the third nucleophile (Nu³) in trapping the presumed sulfenium ion intermediate of the Pummerer rearrangement to complete ring D construction. The absolute configuration at C3, which subsequently influenced the stereochemistry of C15 and C16, was controlled by asymmetric induction of the sulfoxide chirality.

2.3
Diels-Alder Reactions

The Diels-Alder reaction is one of the most useful and versatile reactions in organic synthesis. In a single transformation, two new carbon-carbon bonds and a six-membered cyclic ring system are formed. Numerous efforts have been devoted to the design of new and efficient dienophiles for the Diels-Alder process. Our first demonstration that acetylenic sulfoxides could be used as dienophiles in the Diels-Alder reaction was on the achiral forms [19]. The results are summarized in Table 4. In general, the terminal acetylenic sulfoxide **32a** is more reactive than the methyl substituted acetylenic sulfoxide **32b**.

Later, we embarked on a more systematic study of the Diels-Alder reactions of chiral acetylenic sulfoxides **1a**, **1b** and **1c** (Scheme 9) [20]. The Diels-Alder reactions of the three acetylenic sulfoxides with cyclopentadiene were carried out in appropriate solvents at different temperatures with or without Lewis acid.

Table 4. Diels-Alder reactions of acetylenic sulfoxides

Diene	Acetylenic Sulfoxide	Conditions	Adduct	Yield (%)
	p-NO$_2$-C$_6$H$_4$SOC≡CR			
	32a R = H	r.t./6h/C$_6$H$_6$		97[a]
	32b R = CH$_3$	80°C/6h/C$_6$H$_6$		82[a]
	32a	80°C/2h/C$_6$H$_6$		76[a]
	32a	140°C[b]/7h/C$_6$H$_6$		82
	32b	140°C[b]/20h/C$_6$H$_6$		73
	32a	140°C[b]/8h/C$_6$H$_6$		74[c]
	32b	140°C[b]/20h/C$_6$H$_6$		64[c]
	32a	145°C/25h/Xylene		53
	32a	140°C/1h/Xylene		97
	32b	135°C[b]/12h/C$_6$H$_6$		57
	32a	145°C/2h/Xylene		80[a]

[a] mixture of diastereomers, [b] sealed tube, [c] 1:1 mixture of regioisomers.

Two effects of the Lewis acids were investigated, the reaction rate and the diastereoselectivity. As shown in Table 5, all Lewis acids used enhanced the dienophilicity of the chiral acetylenic sulfoxide and accelerated the reaction. With mild Lewis acids, e.g. ZnCl$_2$, ZnBr$_2$ and MgBr$_2$, the reaction time could be reduced by 60–95%. With stronger Lewis acids, e.g. BF$_3$ · Et$_2$O and TiCl$_4$, the effect on the reaction rate was even more obvious, even in catalytic amounts.

In contrast, the effect of Lewis acids on the diastereoselectivity was disappointing. With all the Lewis acids used, diastereoselectivities were not improved compared to the control experiment. The presence of the electron-withdrawing (**1b**) or electron-donating group (**1c**) on the aromatic ring may have pro-

Table 5. Diels-Alder reactions of chiral acetylenic sulfoxides with cyclopentadiene, Lewis acids effects

Sulfoxide	Solvent	Lewis Acid	Temperature	Time	Diastereoselectivity	Total Yield (%)
1a	CH_2Cl_2 or THF	–	r.t.	32 h	66 : 34	82
1a	CH_2Cl_2 or THF	$ZnCl_2$	r.t.	7 h	63 : 37	80
1a	CH_2Cl_2 or THF	$ZnBr_2$	r.t.	1 h	55 : 45	78
1a	CH_2Cl_2 or THF	$ZnBr_2$	-25°C	8.5 h	57 : 43	71
1a	CH_2Cl_2 or THF	$LiClO_4(s)$	r.t.	9.5 h	64 : 36	78
1b	THF	–	r.t.	20 h	50 : 50	85
1b	THF	$ZnCl_2$	r.t.	8.5 h	50 : 50	81
1b	CH_2Cl_2	$LiClO_4(s)$	r.t.	8.5 h	50 : 50	84
1c	CH_2Cl_2 or THF	–	r.t.	60 h	43 : 57	81
1c	CH_2Cl_2	$ZnBr_2$	r.t.	2.5 h	40 : 60	88
1c	CH_2Cl_2	$ZnBr_2$ (0.15 equiv.)	r.t.	26 h	40 : 60	84
1c	CH_2Cl_2	$MgBr_2$	r.t.	6 h	46 : 54	81
1c	CH_2Cl_2	$LiClO_4(s)$	r.t.	19 h	40 : 60	80
1c	CH_2Cl_2	$BF_3 \cdot Et_2O$ (0.3 equiv.)	r.t.	20 min	46 : 54	81
1c	CH_2Cl_2	$BF_3 \cdot Et_2O$ (0.3 equiv.)	0°C	4 h	42 : 58	75
1c	CH_2Cl_2	$BF_3 \cdot Et_2O$ (0.3 equiv.)	-55°C	7.5 h	40 : 60	78
1c	CH_2Cl_2	$TiCl_4$ (0.3 equiv.)	r.t.	10 min	40 : 60	85
1c	CH_2Cl_2	$TiCl_4$ (0.3 equiv.)	-5°C	3 h	33 : 67	81

vided a second possible coordination site for the Lewis acids, in addition to the sulfoxide oxygen, but this did not improve the results either. A sterically hindered α-methoxy-naphthyl group on the sulfoxide (**1c**), which in other situations greatly improves the diastereoselectivity [11], also did not show any significant effect. The diastereoselectivities were estimated from the well-resolved 270 MHz ^{1}H NMR signals of the cycloadducts. In the case of **1a**, the two diastereomeric cycloadducts could be separated by column chromatography (Scheme 9). The major diastereoisomer **33** $\{[\alpha]_D^{21} = +203.4$ $(c = 2.36, CHCl_3)$; lit. $[\alpha]_D^{25} = +208$; mp 70–71 °C} had been previously transformed by Maignan et al. to optically active $(1R, 4R)$-bicyclo[2.2.1]hept-5-en-2-one (**34**) [21]. Thus, the absolute configuration of the adducts resulting from **1a** could be established from their opti-

1a-c

a : Ar = CH$_3$—◯—

b : Ar = ◯—NO$_2$

c : Ar = ◯◯—OCH$_3$

33

Ar = CH$_3$—◯—

MCPBA

34   **(±)35c**

Scheme 9

cal rotation. However, for **1b** and **1c**, an inseparable mixture of diastereoisomeric cycloadducts resulted. Therefore, their absolute stereochemistry could not be easily established. Nonetheless, when the diastereomeric cycloadducts from **1c** were oxidized with MCPBA, the chirality at the sulfur atom was destroyed to yield a single racemic sulfone (± **35c**) with well defined ^{1}H- and ^{13}C-NMR spectra.

3
Vinyl Sulfoxide

Paquette first demonstrated the reactivity of phenyl vinyl sulfoxide (**36**) in Diels-Alder reactions [22, 23]. We used the vinyl sulfoxide as a two-carbon synthon in the syntheses of alkaloids and heterocycles.

3.1
Alkaloid Synthesis

3.1.1
Hydrohydrastinine

A simple and straightforward application was outlined in the synthesis of hydrohydrastinine as depicted in Scheme 10. Michael addition of 3,4-methylenedioxyphenylmethyl amine to vinyl sulfoxide **36** took place smoothly in refluxing methanol. Pummerer rearrangement in acetic anhydride afforded acetoxysulfide **37** in 90% yield and this was then cyclized to **38** with BF$_3$ etherate in 93% yield. Sulfide **38,** which was rather unstable, was desulfurized with Raney nickel in 80% yield. Hydrolysis of the acetyl group followed by reductive methylation afforded hydrohydrastinine (**39**) in good yield [24].

Scheme 10

3.1.2
Isoquinolone Alkaloids

Isoquinolone alkaloids are a group of naturally occurring alkaloids mainly isolated from *Hernandiaceae* and *Ranunculaceae*. They can be subdivided into two categories: those with a totally aromatic nucleus, such as 6,7-dimethoxy-2-methylisocarbostyril (**43a**) [25] and doryanine (**43b**) [26], and those with a C3–C4 single bond, including *N*-methylcorydaldine (**45a**) and oxyhydrastinine (**45b**) (Scheme 11) [27].

a: R^1 = R^2 = CH_3
b: R^1, R^2 = $-CH_2-$

43a   6,7-dimethoxy-2-methylisocarbostyril
43b   doryanine

45a   *N*-methylcorydaldine
45b   oxyhydrastinine

Scheme 11

Michael addition of methyl amine to phenyl vinyl sulfoxide (36) afforded amine 40 [28]. DCC coupling with a substituted benzoic acid gave 41 in good yield. Pummerer rearrangement of 41 in refluxing acetic anhydride yielded acetoxy sulfide 42 in almost quantitative yield. Treatment of 42 with p-toluenesulfonic acid in refluxing toluene not only effected the cyclization but also the elimination to yield the completed aromatic series 43a and 43b of the isoquinolone alkaloids. This represented a facile convergent route for the total synthesis of 6,7-dimethoxy-2-methylisocarbostyril (43a) and doryanine (43b) in a three-step reaction sequence starting from 40. The overall isolated yields were 70 and 47%, respectively. With a weaker acid, lower reaction temperature and trichloroacetic acid in refluxing benzene, the cyclization product 44 could be isolated. Desulfurization with Raney nickel completed the syntheses of the C3–C4 saturated series N-methylcorydaldine (45a) and oxyhydrastinine (45b).

3.2
Heterocycle Synthesis

3.2.1
Furans and Pyrroles

Efficient syntheses of substituted furans and pyrroles continue to be of interest in view of the widespread occurrence of these systems in nature. Michael addition of β-ketoester 47 to vinyl sulfoxides 36 and 46 proceeded smoothly in the presence of sodium alkoxide (Scheme 12) [29]. It was anticipated that

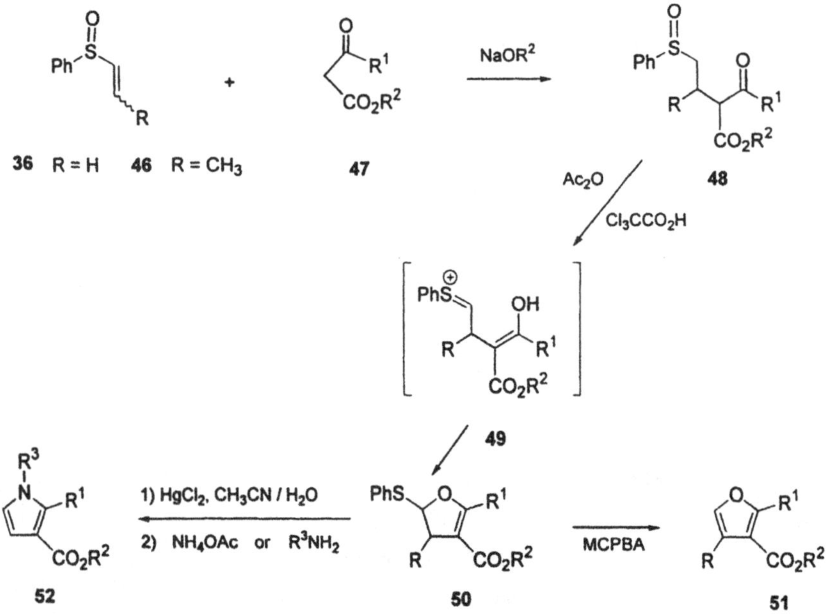

Scheme 12

Pummerer rearrangement of the Michael adducts would produce reactive sulfenium ion intermediates **49** susceptible to a second nucleophile attack; intramolecular trapping by the enol oxygen would then give the cyclized products (**50**). This two-step transformation was achieved directly in good yield by treatment of **48** with trichloroacetic acid and acetic anhydride in refluxing toluene. MCPBA oxidation of sulfide **50** followed by spontaneous elimination afforded furan **51** in good overall yield (Table 6).

Dihydrofuran **50** can be viewed as a latent form of 1,4-dicarbonyl compounds. Replacement of the heterocyclic oxygen atom with an amine nitrogen may provide a synthesis of substituted pyrroles. Several conditions were tried; eventually, it was found that mercury (II) chloride could assist the transformation smoothly. Compound **50** was first refluxed with 2 equiv. of $HgCl_2$ in acetonitrile-water (3:1) for one hour followed by stirring overnight with excess ammonium acetate or primary amines at room temperature. Fair to good yields of the pyrroles were obtained (Table 6) [30].

Table 6. Furan and pyrrole synthesis

R^1	R^2	R	Yield		
			Furan (51)[a]	R^3	Pyrrole (52)[b]
CH_3	C_2H_5	H	46%	H	72%
				$PhCH_2$	60%
				CH_3	60%
				Pr	62%
Ph	C_2H_5	H	50%	H	73%
				CH_3	63%
				$PhCH_2$	40%
				Pr	52%
$PhCH_2$	C_2H_5	H	48%		
C_2H_5	CH_3	H	46%		
CH_3	C_2H_5	CH_3	33%		

[a] from **48**, [b] from **50**

3.2.2
1,3-Dithiole-2-one

1,3-Dithiole-2-one (**60**), which can be readily transformed into its thio- or seleno-carbonyl derivatives, is a key intermediate for the synthesis of tetrathiafulvalene (Scheme 13)[31]. We first anticipated that compound **57**, a Michael addition product of xanthate **54** to vinyl sulfoxide, might be an ideal intermediate for the synthesis of **60** via cyclization under Pummerer rearrangement conditions. However, although Michael addition of dithiocarbamate **53** to vinyl sulfoxide proceeded smoothly to yield compound **55**, the addition reaction with xanthate **54** failed. We then turned to the alkylation approach. Xanthate **54** was alkylated smoothly with **56**, which served as the synthetic equivalent of the vinyl sulfoxide, in ethanol under sonication in 90% yield [32]. Cyclization of **57** under Pummerer rearrangement conditions in the presence of trifluoroacetic acid afforded **58** in 79% yield. Sodium metaperiodate oxidation gave the unstable sulfoxide **59** which underwent thermal elimination to yield **60** in refluxing benzene in moderate yield.

In summary, vinyl sulfoxide (or its equivalent **56**) was adopted as a two-carbon synthon for the syntheses of alkaloids (**39, 43 a b, 45 a b**), furans (**51**), pyrroles (**52**), and 1,3-dithiole-2-one (**60**). Our overall strategy is summarized in Scheme 14. Michael addition of Nu^1 to the vinyl sulfoxide followed by intramolecular trapping of the presumed sulfenium ion Pummerer rearrangement intermedia-

Scheme 13

te **61** by Nu^2 resulted in the cyclic product **62**. In the alkaloid synthesis, the amine nitrogen is Nu^1 and the electron-rich aromatic moiety serves as Nu^2. In the case of furan synthesis, the activated methylene carbon is Nu^1 while Nu^2 is the enol oxygen. Two possible routes to further transform **62** to the final targets were explored. Direct desulfurization gave **63** (i.e. **39** and **45ab**) and oxidation followed by sulfoxide elimination afforded **64** (i.e. **43ab**, **51** and **60**). In this regard, the vinyl sulfoxide **36** can be viewed as an alkyl or alkenyl 1,2-dielectrophilic two-carbon synthon for structures **65** and **66**, respectively.

Scheme 14

4
Acetylenic Sulfinate and Sulfonate

Sulfinate and sulfonate are important functional groups in organic synthesis [1b]. For example, sulfinates are key starting materials for sulfoxide syntheses and sulfonates have been extensively used as leaving groups via the cleavage of the C–O bond. However, their ability, as electron-withdrawing groups, to activate an unsaturated carbon unit have not yet been fully explored, with the exception of vinylic sulfonates which have been successfully used as dienophiles in both intermolecular [33] and intramolecular [34] Diels-Alder reactions. To follow our studies on acetylenic sulfoxides, we prepared the previously unknown acetylenic sulfinate and sulfonate and explored their reactivity as dienophiles in Diels-Alder reactions.

4.1
Preparation

The preparation of acetylenic sulfinates **68a–c** was accomplished in a two-step one-pot reaction sequence (Scheme 15). At –20 °C, in the presence of a large excess of thionyl chloride, cyclohexanol was converted into cyclohexyl chlorosulfinate (**67**) [35]. After removal of the excess thionyl chloride at 0 °C in vacuo, the labile chlorosulfinate was treated with the corresponding lithium acetylide to afford high yields of **68a–c**. Desilylation of **68a** or **68b** to the parent acetylenic sulfinate **68d** [36] was achieved by treatment with potassium fluoride in acetonitrile.

Scheme 15

The first synthesis of acetylenic sulfonate **72** was achieved by MCPBA oxidation of sulfinate **68a** to **71**, followed by potassium fluoride desilylation, in 90% overall yield. However, to our surprise, the triisopropylsilyl-protected acetylenic sulfinate **68b** resisted oxidation with the oxidants we tried, including MCPBA, oxone, H_2O_2/SeO_2, and RuO_4 generated from $RuCl_3/NaIO_4$ [37].

4.2
Diels-Alder Reactions

The results of the Diels-Alder reactions of acetylenic sulfinates **68a**, **68c** and **68d** with a series of dienes are summarized in Table 7. For a reactive diene, such as cyclopentadiene, the cycloaddition took place readily at room temperature to afford excellent yields of adducts for all three sulfinates. Apparently, the steric hindrance at the terminal acetylenic position hardly hindered the reaction in contrast to the reaction with acetylenic sulfoxides. By virtue of the asymmetric center at the sulfinate group and the dissymmetry element present in the substituted [2.2.1] bicyclic ring system, the Diels-Alder adducts were formed as mixtures of diastereoisomers which were inseparable. Mild oxidation of the adduct by MCPBA, thus eliminating the chirality at the sulfur group, afforded the corresponding sulfonates as single diastereomers. For less reactive dienes, the Diels-Alder reaction was carried out at elevated temperature. With an unsymmetrical diene, such as isoprene, a 3:2 ratio of regioisomers was obtained. In the case of the reaction between **68d** and 1-trimethylsiloxybutadiene, loss of the trimethylsiloxy group with concomitant aromatization was observed.

In order to obtain an insight into the diastereoselectivity in the Diels-Alder reaction of acetylenic sulfinates, chiral (+)-*trans*-2-phenylcyclohexanol [35] was used in place of cyclohexanol in the synthesis of the dienophile. A 1:1 diastereoisomeric mixture of acetylenic sulfinates **69** and **70** was obtained. After separation, each diastereoisomer was subjected to a Diels-Alder reaction with cyclopentadiene. Although the reaction once again occurred readily at room temperature, to our disappointment an inseparable mixture of diastereomeric adducts (3:2 by NMR) was obtained for each sulfinate. Apparently, a more spatially demanding chiral auxiliary needs to be incorporated into the dienophile in order to generate chiral sulfinates which cycloadd with prominent diastereoselectivity.

Acetylenic sulfonate **72** is relatively less stable and has to be stored at 0°C to avoid decomposition. The results of the Diels-Alder reaction of **72** are summarized in Table 8 [38]. For a reactive diene, such as cyclopentadiene, the cycloaddition took place readily at 0°C. For less reactive dienes, the reactions were carried out at elevated temperature. Mixtures of regioisomers resulted when unsymmetrical dienes were used. Sulfonate is a powerful electron-withdrawing group. Among the three acetylenic dienophiles (sulfoxide **1**, sulfinate **68d** and sulfonate **72**) we studied, the acetylenic sulfonate is the most reactive one.

Table 7. Diels-Alder reactions of acetylenic sulfinates

Diene	Dienophile	Conditions T (°C) t (h)	Adduct	Yield (%)
(cyclopentadiene)	**68a**	25/CH$_2$Cl$_2$ 10	C$_6$H$_{11}$O–S(=O)– ; (CH$_3$)$_3$Si	90[a]
	68d	25/CH$_2$Cl$_2$ 5	C$_6$H$_{11}$O–S(=O)– ; H	95[a]
	68c	25/CH$_2$Cl$_2$ 8	C$_6$H$_{11}$O–S(=O)– ; Bu	90[a]
(1,3-cyclohexadiene)	**68d**	60/C$_6$H$_6$ 8	C$_6$H$_{11}$O–S(=O)– ; H	86[a]
(isoprene)	**68a**	50[c]/C$_6$H$_6$ 24	C$_6$H$_{11}$O–S(=O)– ; CH$_3$; (CH$_3$)$_3$Si	76[b]
	68d	50[c]/C$_6$H$_6$ 12	C$_6$H$_{11}$O–S(=O)– ; CH$_3$; H	81[b]
(2,3-dimethylbutadiene)	**68d**	60/C$_6$H$_6$ 12	C$_6$H$_{11}$O–S(=O)– ; H	84
(1-trimethylsilyloxybutadiene), O–Si(CH$_3$)$_3$	**68d**	130[c]/C$_6$H$_6$ 6	C$_6$H$_{11}$O–S(=O)–	95

[a] as a mixture of diastereomers,   [b] 3:2 mixture of regioisomers,   [c] sealed tube.

Table 8. Diels-Alder reactions of acetylenic sulfonate 72

Diene	Conditions		Adduct	Yield (%)
	T (°C)	t (h)		
	$0/CH_2Cl_2$	8	 SO_3R $R = C_6H_{11}$	99
	$80/C_6H_6$	30	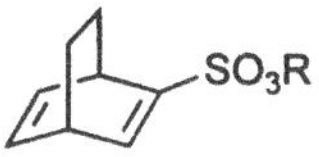 SO_3R	94
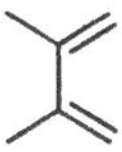	$60^a/C_6H_6$	9	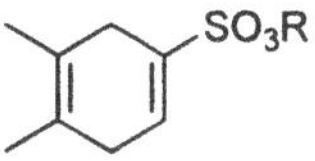 SO_3R	65
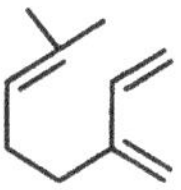	$80/C_6H_6$	11	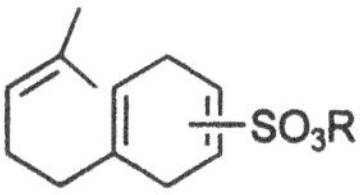 SO_3R	78[b]
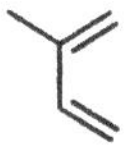	$50^a/C_6H_6$	37	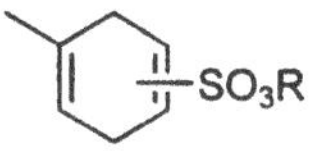 SO_3R	57[b]
	$50^a/CH_2Cl_2$	60	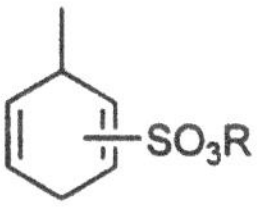 SO_3R	64[b]

[a] sealed tube, [b] mixture of regioisomers.

5
α,β-Unsaturated Propane Sultone (1-Propene-1,3-Sultone)

Although the chemistry of the saturated propane sultone was investigated in some detail [39], there were only limited reports on the preparation and reaction of the corresponding α,β-unsaturated propane sultone.

5.1
Preparation

Our approach to the synthesis of the unsaturated γ-sultone 5 is depicted in Scheme 16. Sodium 2-propenesulfonate (73) was prepared from allyl bromide and sodium sulfite. Bromination of 73 in water resulted in dibromide 74. Distillative cyclization of the dibromide under acidic conditions afforded β-bromo-sultone 75 which could be eliminated to 5 upon treatment with triethylamine.

73

74

75

75

m.p. 81-83°C

Scheme 16

5.2
Diels-Alder Reactions

Vinyl sulfonates were found to be reactive dienophiles in both intermolecular and intramolecular Diels-Alder reactions [33, 34]. The results of the Diels-Alder reaction of 5 with various dienes are summarized in Table 9. The reaction yields are high and the *endo/exo* selectivities for cyclic dienes are reasonably good [40].

5.3
Ring Opening of Cycloadducts and Synthesis of Chiral Sultams

The sultone cycloadducts could be further manipulated by ring-opening with various nucleophiles, such as alcohols and amines, at the γ-position [41]. When optically active (S)-$(-)$-α-methylbenzylamine reacted with the racemic sultone cycloadduct **76** in ethanol at room temperature, one of the diastereomeric ammonium sulfonates precipitated from the reaction mixture (Scheme 17). Although the absolute stereochemistry of **77** had not been determined, cyclization of optically pure **77** with phosphorus oxychloride gave an optically pure sultam **78**. Formic acid debenzylation followed by base hydrolysis of the *N*-formyl group afforded the optically pure sultam **80** in good yield [40].

Optically pure sultams have been used by Oppolzer as chiral auxiliaries in various asymmetric transformations, including Diels-Alder reaction, aldolization, conjugate addition, *bis*-hydroxylation, and catalytic hydrogenation [42, 43]. In the literature, the most commonly used chiral sultam is derived from camphor (Oppolzer's sultam). The ready access to **80** and other chiral sultams from the Diels-Alder cycloadducts could further expand the scope of their use as chiral auxiliaries in asymmetric synthesis.

Table 9. Diels-Alder reactions of α, β unsaturated propane sultone 75

Diene	Conditions		Adduct	Yield (%)
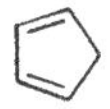	T (°C)	t (h)	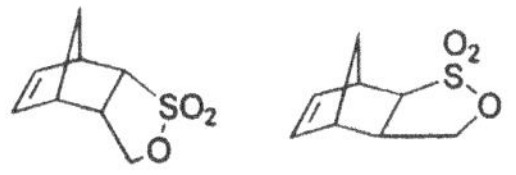	
	20/CH$_2$Cl$_2$	7d	84 : 16	98
	120[a]/Toluene	4	73 : 27	96
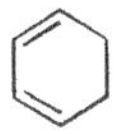	150[a]/Toluene	18	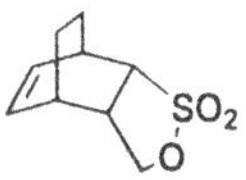 *endo* only	96
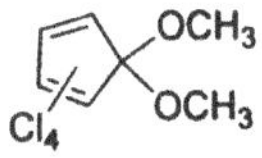	140/Xylene	20	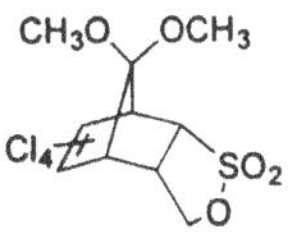 *endo* only	72
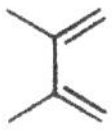	150[a]/Toluene	18	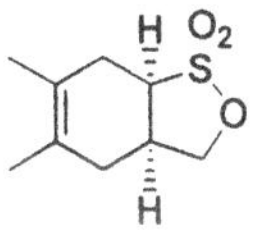	96
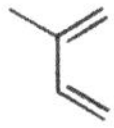	140[a]/Toluene	20	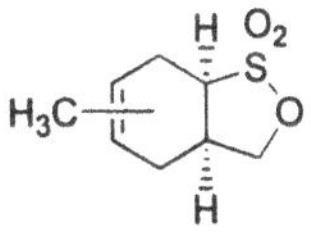	84[b]
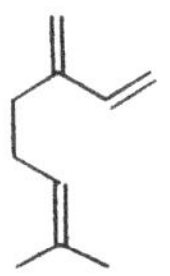	140/Xylene	20	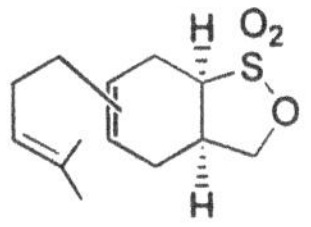	75[b]

[a] sealed tube, [b] mixture of regioisomers.

Scheme 17

Acknowledgements. We thank our coworkers, whose names appear in the references, for their crucial contributions which made this work possible. Financial support received from the Research Grant Council (HKBC 109/93E and HKBC 136/94P) and the Faculty Research Grant (FRG/95-96/II-29) is gratefully acknowledged.

References

1. a) Patai S, Rappoport Z, Stirling CJM (eds) (1983) The chemistry of sulphones and sulphoxides. John Wiley, New York
   b) Patai S, Rappoport Z (eds) (1991) The chemistry of sulphonic acids, esters and their derivatives. John Wiley, New York
2. a) Solladie G (1983) Addition of chiral nucleophiles to aldehydes and ketones. In: Morrison JD (ed) Asymmetric synthesis, vol 2. Academic, New York p 157
   b) Walker AJ (1992) Tetrahedron Asymmetry 3:961
3. Carreno MC (1995) Chem Rev 95:1717
4. Pitchen P, Dunach E, Deshmukh NN, Kagan HB (1984) J Am Chem Soc 106:8188
5. Furia D, Modena G, Seraglia R (1984) Synthesis 325
6. a) Andersen KK (1962) Tetrahedron Lett 1962:93
   b) Anderson KK, Gaffield W, Papanikolau NE, Foley JW, Perkins RI (1964) J Am Chem Soc 86:5637

7. Lee AWM, Chan WH, Lee YK (1991) Tetrahedron Lett 32:6861
8. Lee AWM, Chan WH, Tao Y, Lee YK (1994) J Chem Soc Perkin Trans 1 477
9. Klunder JM, Sharpless KB (1987) J Org Chem 52:2598
10. a) Pyne SG, Bloem P, Chapman SL, Dixon CE, Griffith R (1990) J Org Chem 55:1086
    b) Pyne SG (1987) Tetrahedron Lett 28:4737
11. Pyne SG, Hajipour AR, Prabakaran K (1994) Tetrahedron Lett 34:6481
12. Chan WH, Lee AWM, Jiang L (1995) Tetrahedron Lett 36:715
13. a) Kutney JP (1977) The synthesis of indole alkaloids. In: ApSimon J (ed) The total synthesis of natural products, vol 3. John Wiley, New York, p 273
    b) Baxter EW, Mariano PS (1992) Recent advances in synthesis of yohimbine alkaloids. In: Pelletier SW (ed) Alkaloids: chemical and biological perspectives, vol 8, Springer, Berlin Heidelberg New York, p 197
14. Aube J, Ghosh S, Tanol M (1994) J Am Chem Soc 116:9009
15. Craig D, Deniels K (1992) Tetrahedron 48:7803
16. Lee AWM, Chan WH, Mo T (1996) Chiral acetylenic sulfoxides in enantioselective synthesis: Asymmetric synthesis of pentacyclic yohimbine alkaloids. Presented at the 212th American Chemical Society National Meeting, Orlando
17. Okamura K, Yamada S (1978) Chem Pharm Bull 26:2305
18. Chatterjee A (1986) Pure & Appl Chem 58:685
19. Lee AWM, Chan WH, Wong MS (1988) J Chem Soc Chem Commun 1585
20. Lee AWM, Chan WH, Ji FY, Poon WH (1995) J Chem Research (S) 368
21. Maignan C, Belkasmioui F (1988) Tetrahedron Lett 29:2823
22. Carr RVC, Paquette LA (1980) J Am Chem Soc 102:853
23. Paquette LA, Carr RVC (1985) Phenyl vinyl sulfone and sulfoxide. In: Kende AS (ed) Organic synthesis, vol 64. Wiley, New York. p 157
24. Lee AWM, Chan WH, Chung TS, Wong JCS, unpublished results
25. Mollov NM, Dutschewska HB (1969) Tetrahedron Lett 1951
26. Belgaonkar VH, Usgaonkar RN (1977) J Chem Soc Perkin Trans 1 702
27. Shamma M, Podczazy MA (1971) Tetrahedron 27:727
28. Lee AWM, Chan WH, Chan ETT (1992) J Chem Soc Perkin Trans 1 309
29. Lee AWM, Chan WH, Chan ETT (1992) J Chem Soc Perkin Trans 1 945
30. Chan WH, Lee AWM, Lee KM, Lee TY (1994) J Chem Soc Perkin Trans 1 2355
31. Krief A (1986) Tetrahedron 42:1209
32. Lee AWM, Lee YK, unpublished results
33. a) Distler H (1965) Angew Chem Int Ed Engl 4:300
    b) Klein LL, Deeb TM (1985) Tetrahedron Lett 26:3935
34. Metz P, Fleischer M, Fröhlich R (1995) Tetrahedron 51:711 and references therein
35. Whitesell JK (1992) Chem Rev 92:953
36. Chan WH, Lee AWM, Lee KM (1994) J Chem Research (S) 138
37. Gao Y, Sharpless KB (1988) J Am Chem Soc 110:7538
38. Lee AWM, Chan WH, Zhong ZP, Lee KF, Yeung ABW unpublished results
39. Roberts DW, Williams DL (1987) Tetrahedron 43:1027
40. Lee AWM, Chan WH, Jiang LS (1996) Chemistry of α, β unsaturated γ-sultone: Diels-Alder reactions and synthesis of an optically pure sultam. Presented at the 212th American Chemical Society National Meeting, Orlando
41. Buglass AJ, Tillett JG (1991) Sultones and sultams In: Patai S, Rappoport Z (eds). The chemistry of sulphonic acids, esters and their derivatibes. Wiley, New York, p 789
42. Oppolzer W (1981) Tetrahedron 43:1969
43. Oppolzer W (1990) Pure Appl Chem 62:1241

N-Sulfonyl Imines – Useful Synthons in Stereoselective Organic Synthesis

Steven M. Weinreb

Department of Chemistry, The Pennsylvania State University, University Park, PA 16802 USA, e-mail: smw@chem.psu.edu

Until recently, *N*-sulfonyl imines had found only limited and sporadic use in organic synthesis. During the past decade, however, it has become increasingly clear that these species are valuable synthons and are capable of undergoing many unique transformations. A comprehensive review of the chemistry of these compounds is presented here with particular emphasis on their applications in stereoselective processes. Methods for preparing *N*-sulfonyl imines are outlined, along with a survey of their uses in a wide range of addition, pericyclic and cycloaddition reactions.

Keywords. Sulfonamides, pericyclic reactions, [2+2]cycloadditions, [4+2]cycloadditions, ene reactions, nucleophilic additions

Table of Contents

Topics in Current Chemistry, Vol. 190
© Springer Verlag Berlin Heidelberg 1997

1
Introduction

Electron-deficient imines and iminium complexes are now generally accepted as valuable intermediates in the construction of a variety of nitrogen-containing molecules. In particular, N-acyl imines have become widely recognized as versatile synthons [1]. These species undergo a diverse array of synthetically useful reactions including various types of cycloaddition, nucleophilic addition and amidoalkylation. Interestingly, the analogous electron-deficient N-sulfonyl imines have received far less attention, and only during the past few years has the real potential of this functionality begun to emerge. One of the major reasons why N-sulfonyl imines have not been very widely utilized to date may have been a lack of reliable and general methods for generating these compounds. However, as is described below there has been significant remediation of this problem in recent years. It might be noted that N-sulfonyl imines appear to have the high reactivity characteristic of N-acyl imines. However, N-sulfonyl imines can often be isolated, and are reasonably stable compounds, whereas N-acyl imines usually undergo rapid oligomerization and are rarely observed [1]. In addition, N-sulfonyl imines often undergo reactions which do not occur with more common N-alkyl and N-aryl imines.

This article outlines the methodology currently available for producing N-sulfonyl imines. In addition, a survey of the applications of this functionality in organic chemistry is presented, with particular emphasis on applications to stereoselective synthesis.

2
Preparation of *N*-Sulfonyl Imines

The methods used for generating *N*-sulfonyl imines were very slow to develop prior to the burst of interest in this area during the past ten years. *N*-Sulfonyl imines can often be produced in situ from more stable precursors such as α-alkoxy- or α-hydroxy sulfonamides. However, a number of good procedures now exist for direct synthesis of *N*-sulfonyl imines, particularly those derived from non-enolizable aldehydes. It might be noted that there is still a lack of good procedures for synthesizing *N*-sulfonyl imines from enolizable aldehydes and ketones. The section below outlines the primary methods known for generating *N*-sulfonyl imines, although some additional scattered experimental variations of these methods do exist.

2.1
Direct Formation from Primary Sulfonamides and Aldehydes/Ketones/Acetals

The earliest procedure described for synthesis and isolation of *N*-sulfonyl imines of aryl aldehydes utilized $ZnCl_2$ as a catalyst [Eq. (1)] [2]. Yields generally ranged from ~20–70% of crystalline products. A related procedure published shortly thereafter by a Russian group using $AlCl_3$ seems to produce somewhat higher isolated yields of the aryl *N*-sulfonyl aldimines [3]. More recently, an improved and milder variation of this type of condensation was described by Jennings and Lovely [4]. Thus, aromatic aldehydes could be combined with primary sulfonamides using titanium tetrachloride/triethyl amine at 0 °C. Isolated yields of imine here generally ranged from 50–70% [Eq. (2)]. Although the procedure is not useful for preparing *N*-sulfonyl imines from enolizable aldehydes and ketones (presumably due to competing aldol reactions), the *N*-tosyl imine of

$$ArSO_2NH_2 \quad + \quad Ar'CHO \quad \xrightarrow[\quad]{\substack{ZnCl_2 \\ \text{or} \\ AlCl_3}} \quad Ar' \text{—CH=N—SO}_2Ar \qquad (1)$$

Ar=Ph, *p*ClPh   Ar'=Ph, *m*NO$_2$Ph,
                 *p*Me$_2$NPh, 2-furyl

$$RSO_2NH_2 \quad + \quad Ar'CHO \quad \xrightarrow[\quad 50\text{-}70\% \quad]{\substack{TiCl_4/CH_2Cl_2 \\ NEt_3/0\ ^\circ C}} \quad Ar' \text{—CH=N—SO}_2R \qquad (2)$$

R=*p*MePh, Me   Ar'=Ph, *m*NO$_2$Ph,
                 *p*MeOPh, 2-naphthyl

(+)-camphor could be synthesized in moderate yield [Eq. (3)]. Davis and co-workers have independently used a similar methodology to prepare closely related camphor sulfonamides [5a], but in better yields [Eq. (4)]. Interestingly, these camphor-derived imines are apparently chromatographically stable, although those produced from aldehydes are not [4].

$$\text{(3)}$$

TsNH$_2$
TiCl$_4$/NEt$_3$
PhMe/Δ
34%

NTs

$$\text{(4)}$$

RCH$_2$SO$_2$NH$_2$
TiCl$_4$/NEt$_3$
C$_2$H$_3$Cl$_3$/Δ

NSO$_2$CH$_2$R
R=Me (70%)
R=Ph (64%)

Another method which has been utilized for condensing benzaldehyde and *p*-toluenesulfonamide involves azeotropic distillation of water in the presence of an acid ion exchange resin and 4Å molecular sieves [5b]. The crystalline *N*-sulfonyl aldimine could be isolated in 87% yield on a 200g scale.

Kresze and coworkers have reported that simply heating a neat mixture of an aryl sulfonamide and an ethyl or methyl acetal from an aromatic aldehyde affords the *N*-sulfonyl imine in good yields [6] [Eq. (5)]. However, with the diethyl acetal of ethyl glyoxylate, only the bis-sulfonamido acetal was produced. No indication was given if this procedure was attempted with acetals of aliphatic aldehydes.

$$\text{ArSO}_2\text{NH}_2 \quad + \quad \text{Ar'CH(OEt)}_2 \xrightarrow[56\text{-}96\%]{\substack{\text{neat} \\ 150\,^{\circ}\text{C}}} \text{Ar'CH=N}{-}\text{SO}_2\text{Ar} \qquad \text{(5)}$$

Ar=*p*MePh, Ph   Ar'=Ph, *m*NO$_2$Ph,
*p*NO$_2$Ph, *p*ClPh   *p*MeOPh, *p*NO$_2$Ph

2.2
Use of "Activated" Sulfonamides

Kresze pioneered the use of *N*-sulfinyl sulfonamides in the generation of *N*-sulfonyl imines [6–8]. The *N*-sulfinyl sulfonamides **1** are generally readily produced from the parent sulfonamide and thionyl chloride [8, 9] and can be isolated, but are often used as formed in situ (Scheme 1). It was found that a variety of non-enolizable aldehydes are converted to the *N*-sulfonyl imines by heating in benzene with the *N*-sulfinyl sulfonamides 1 in the presence of a catalytic amount of aluminum chloride.

Scheme 1

The reaction probably involves an initial [2 + 2]-cycloaddition of the aldehyde and the *N*-sulfinyl compound to produce an adduct 2 [10] which loses sulfur dioxide to yield the *N*-sulfonyl imine. Isolated yields of the sulfonyl imines shown in Scheme 1 were generally quite good. In the case of the enolizable aldehyde dichloroacetaldehyde, only a low yield of *N*-sulfonyl imine was produced. An attempt was also made using these same reaction conditions to convert butyraldehyde to the corresponding *N*-tosyl imine [6]. However, all that could be isolated here was the bis-sulfonamido acetal.

More recently, in a series of papers Weinreb and coworkers have found that *N*-sulfonyl aldimines can in fact be rapidly produced in situ from aliphatic aldehydes using *N*-sulfinyl sulfonamides and boron trifluoride etherate as catalyst at low temperature [11–15] [Eq. (6)]. Similarly, aliphatic aldehydes can be converted to the *N*-sulfonyl imines in the absence of a Lewis acid at room temperature or above, but more slowly. The former procedure also works well for aromatic aldehydes, whereas the reaction is too slow to be useful if a Lewis acid is not used. The *N*-sulfonyl imines generated in this manner can be utilized in a number of transformations (vide infra). However, except in rare cases [16], *N*-sulfonyl ketimines cannot be formed by this methodology.

In a related method, Trost and Marrs found that the bis-imido tellurium reagent **3**, generated in situ from tellurium metal and chloramine T, reacts with a wide variety of aromatic conjugated and aliphatic aldehydes in refluxing toluene to afford the corresponding *N*-tosyl imines in excellent yields (Scheme 2) [17]. The transformation is thought to occur via tellurocycle **4**, which collapses to the imine and intermediate **5**. From the stoichiometry of the reaction it appears **5** is capable of converting an aldehyde to the *N*-sulfonyl aldimine as effectively as bis-imide **3**. The final inorganic product of the reaction is TeO_2.

Scheme 2

2.3
From Oximes

Studies by Hudson and coworkers have demonstrated that both *N*-sulfonyl aldimines and ketimines can be prepared from the corresponding aldoxime or ketoxime [18]. Thus, treatment of the oxime with a sulfinyl chloride initially affords the *O*-sulfinylated oxime **6** (Scheme 3). If **6** is warmed, it rearranges via a free radical process into an *N*-sulfonyl imine. This transformation appears to provide one of the best and most general routes to *N*-sulfonyl imines and has been extensively exploited recently by Boger and coworkers [19] in hetero Diels-Alder reactions (see Section 5.2).

R, R'=Ph, Me, *p*MePh   R"=Me, *p*MePh
   CHPh₂

Scheme 3

Boger and Corbett have also recently described a convenient modification of the original Hudson methodology [20]. Their procedure is based upon the known [21] propensity of methanesulfonyl- and toluenesulfonyl cyanide to rearrange to the corresponding sulfinyl cyanate (cf. **8**, Scheme 4). Thus, treatment of an oxime with commercially available tosyl cyanide (**7**) generates **8** in situ, which leads to the *O*-sulfinylated oxime **9** and then to the *N*-tosyl imine. This methodology avoids the use of reactive, often unstable sulfinyl chlorides.

Scheme 4

2.4
Sulfonylation of Imines and *N*-Silyl Imines

The direct *N*-sulfonylation of simple NH imines has not been studied to any significant degree despite the availability [22] of these precursors. In a rare use of this approach, Hudson and coworkers [18] have reported two examples of *N*-sulfonylation of ditolyl imine (**10**) to afford the *N*-sulfonyl imines [Eq. (7)] in reasonable yields.

$$\text{(7)}$$

More recently, Georg et al. [23] have found that *N*-trimethylsilyl imines **11** of aromatic non-enolizable aldehydes and ketones, prepared by the methodology of Hart [24], can be converted to the corresponding *N*-sulfonyl imines using aryl or alkyl sulfonyl chlorides (Scheme 5). Unfortunately, the procedure is not applicable to forming *N*-sulfonyl imines from enolizable aldehydes and ketones.

Scheme 5

2.5
Oxidation of Sulfonamides

Shono and coworkers have examined the electrochemical oxidation of sulfonamides [25], presumably with the intent of generating α-alkoxy sulfonamides. However, anodic oxidation of short chain acyclic sulfonamides, like **12**, in the presence of halide ion surprisingly afforded the α-sulfonamido acetals **13** (Scheme 6) [25a]. It is believed that oxidation of **12** occurs to initially produce α-methoxy sulfonamide **14**. Under the reaction conditions, however, **14** eliminates methanol to produce *N*-sulfonyl aldimine **15**, which can tautomerize to ene sulfonamide **16**. Reaction of **16** with a positive halogen species, generated electrochemically, probably leads to **17**, which can rearrange via an intermediate aziridine to the observed acetal product **13**.

Scheme 6

When longer chain sulfonamides, such as **18**, were employed in the oxidation, a mixture of α-sulfonamido acetal **19** and pyrrolidine **20** was produced [Eq. (8)]. It was postulated that **20** is formed via a free radical process of the Hofmann-Loffler type, involving a 1,5-hydrogen atom transfer. It might be noted that α-methoxy sulfonamide **14** (R = Et) could in fact be observed spectroscopically at low temperature.

This electrochemical methodology has also been applied to oxidation of *N*-tosyl azetidines [25b]. Thus, anodic oxidation of sulfonamide **21** in acetic acid yielded α-acetoxy sulfonamide **22** [Eq. (9)]. Such compounds are useful precursors of *N*-sulfonyl iminium ions (vide infra). No indication was given, however, as to whether this procedure can be extended to other ring systems.

Han and Weinreb [26] have attempted to develop a non-electrochemical approach towards α-oxidation of sulfonamides based upon the earlier work of Pines et al. [27]. The strategy here was to expose an *o*-diazo arylsulfon-

amide **23** to a catalytic amount of cuprous ion to produce aryl radical **24**, which would undergo 1,5-hydrogen atom transfer [28] to yield a new radical **25** (Scheme 7). Cu^{2+}-promoted oxidation of **25** would then afford the *N*-sulfonyl iminium species **26**, which should add solvent to yield α-alkoxy sulfonamide **27**.

In a test of this approach, the readily available *o*-nitro sulfonamide **28** derived from pyrrolidine was reduced to amine **29** [Eq. (10)]. Exposure of this compound

Scheme 7

to the conditions shown led to the desired α-methoxy sulfonamide **30** in good yield. Unfortunately, the extension of this procedure to other ring systems has, to date, been disappointing. Amino sulfonamide **31** produced a mixture of α-methoxy sulfonamide **32** and the reduced product **33** resulting from hydrogen atom abstraction, perhaps from solvent, by the aryl radical formed initially [Eq. (11)] (cf **24**). Similarly, the seven-membered ring system **34** yielded a mixture of the three products shown in Eq. (12).

2.6
From Reduction of *N*-Sulfonyl Lactams

A convenient approach to α-hydroxy sulfonamides involves hydride reduction of *N*-sulfonyl lactams. For example, Ahman and Somfai [29] have described DIBALH reduction of simple 5- and 6-membered *N*-tosyl lactams to the corresponding α-hydroxy sulfonamides (Scheme 8). It was possible to convert these

Scheme 8

compounds to the α-methoxy sulfonamides via the *N*-sulfonyl iminium species. Both hydroxy compounds and methyl ethers of this type are effective *N*-sulfonyl iminium ion precursors (see Section 4). Interestingly, the corresponding 7-membered *N*-tosyl lactam afforded a complex mixture of products upon similar reduction.

3
Structural Considerations

Relatively little information on the structure of *N*-sulfonyl imines is currently available. Two groups have studied the *E/Z*-isomerization of *N*-sulfonyl imines by NMR methods [30, 31]. The barriers to *E/Z* interconversion of these imines is quite low relative to the related oximes. This phenomenon has been ascribed to a nitrogen inversion mechanism involving $(p-d)$ π conjugation between sulfur and nitrogen which stabilizes the transition state for stereomutation [32, 33]. On the NMR time scale at room temperature one cannot detect geometrical isomers of unsymmetrical *N*-sulfonyl imines [18].

N-Sulfonyl imines derived from enolizable aldehydes and ketones are, in principle, capable of tautomerization to the corresponding ene sulfonamides. There has been no systematic study of this process, probably due in large part to the fact that only relatively few sulfonyl imines of this type have to date been prepared and characterized. Trost and Marrs [17], however, have found that aldehyde **35** on conversion to imine **36** using tellurium reagent **3** led to enamide **37** upon workup (Scheme 9). Similarly, aldehyde **38** was converted to imine **39** which tautomerized to **40** upon isolation.

Scheme 9

4
Nucleophilic Additions

4.1
Hetero Nucleophiles

Not surprisingly, *N*-sulfonyl imines, which are highly electrophilic species, are quite prone to hydrolysis. Thus, addition of water to imine **41** initially produces a "methylol" derivative **42** which in certain cases can be reasonably stable ($R = CCl_3$, CO_2R) [2, 6]. However, this type of intermediate usually dissociates to an aldehyde and a primary sulfonamide (Scheme 10).

Scheme 10

Scattered examples exist of additions of other hetero nucleophiles to *N*-sulfonyl imines. For example, Kresze and coworkers found that thiols add to imine **43** to afford adducts **44** in good yields [Eq. (13)] [6, 34]. Similarly, aniline adds to chloral-derived *N*-sulfonyl imine **46** to afford **45**, and ethanol adds to produce **47** (Scheme 11) [6].

Scheme 11

In a unique example of the application of a phosphorous nucleophile, Shono and coworkers [25b] described addition of trimethyl phosphite to α-acetoxy sulfonamide **22** in the presence of a Lewis acid [Eq. (14)]. The product **49** is presumably formed via nucleophilic phosphite addition to an intermediate *N*-sulfonyl iminium ion **48** in an Arbuzov-like reaction.

$$(14)$$

4.2
Cyanide

Condensation of an alkyl or aryl sulfonamide, formaldehyde or acetaldehyde and potassium cyanide to produce adducts **50** was described a number of years ago in a patent [Eq. (15)] [35]. It is possible that this transformation occurs by addition of cyanide to an intermediate *N*-sulfonyl imine.

Ahman and Somfai [Eq. (16)] have utilized the α-hydroxy and α-methoxy sulfonamides **51** (cf Scheme 8) in reactions with trimethylsilyl cyanide and Lewis acids [29]. It was found that stannic chloride was more effective than

$$(15)$$

$$(16)$$

titanium tetrachloride and boron trifluoride etherate in promoting conversion of compounds **51** to nitriles **53**, presumably via *N*-sulfonyl iminium species **52**. This methodology was also applied to an enantioselective synthesis of the C_2-symmetric amine **57** (Scheme 12). *N*-Tosyl lactam **54** was prepared from L-pyroglutamic acid and was reduced to a 3:1 mixture of α-hydroxy sulfonamides **55**. Exposure of **55** to TMSCN and stannic chloride gave a high yield of a single nitrile **56** with the *trans* configuration. This compound could then be converted into amine **57** in four steps.

Two additional examples of additions of cyanide to an in situ generated *N*-sulfonyl iminium ion are shown in Eqs. (17) and (18). Thus, treatment of azetidine derivative **22** with TMSCN and titanium tetrachloride led to nitrile **58** [25b]. Similarly, the various α-oxygenated sulfonamides in Eq. (18) were converted to the nitrile [34]. In general, the acetate and methyl ether gave the best yields of nitrile using $TiCl_4$ and $SnCl_4$. Lower yields of nitrile were obtained using the α-hydroxy sulfonamide and with BF_3 etherate and ZnI_2 as the Lewis acids.

$$\text{Scheme 12}$$

$$\text{(17)}$$

$$\text{(18)}$$

R=H, Me, Ac

4.3
Stabilized Carbanions and Enol Derivatives

Few examples of additions of stabilized carbanions to *N*-sulfonyl imines currently exist. Kresze et al. [6] have added sodio diethyl malonate to the benzaldehyde/*p*-toluenesulfonamide *N*-sulfonyl imine to produce adduct **59** [Eq. (19)].

$$\text{(19)}$$

More recently, Yamamoto and coworkers [36] have developed a new acyl anion equivalent based upon the ethoxyethyl-protected α-hydroxymalonodinitrile derivative shown in Scheme 13 and have applied it in the area of *N*-sulfonyl imine chemistry. Thus, the carbanion derived from the dinitrile was added to imine **60** to afford adduct **61** which was rather unstable. However, *N*-alkylation of **61** with chloromethyl methyl ether yielded the stable product **62**. Removal of

Scheme 13

Scheme 14

the ethoxyethyl protecting group of **62** led to the unstable dicyano alcohol **63**, which on treatment with glycine methyl ester afforded amide **64**, probably via an intermediate acyl cyanide.

Saigo and coworkers have recently investigated the stereochemistry of the addition of silyl ketene acetals to *N*-sulfonyl imines [37]. In preliminary studies, the addition of *E/Z* mixtures of ketene silyl acetal **66** to an *N*-sulfonyl imine was investigated (Scheme 14). From these results, it appears that the major product of the reaction is always the *anti* isomer **67**, and that the products **67** and **68** are both derived from the *E* isomer of **66** (ie, the *Z* isomer is totally unreactive). Since the ketene acetal **66** configuration was critical to the condensation reaction, the bis-silyl compounds **70** were investigated as an alternative (Scheme 15). It was found that this type of ketene acetal reacts stereoselectively with *N*-sulfonyl imines **69** using TiBr$_4$ as catalyst. As can be seen from the data in Scheme 15, the *anti* products **71** predominated over the *syn* **72** in all cases.

These authors have rationalized the stereochemical results by assuming that Lewis acid coordination of the *N*-sulfonyl imine **69** occurs at the sulfonyl oxygen, thereby producing eight-membered ring transition states for the condensation (Fig. 1). It appears that transition state **74**, leading to the *syn* products **72**, is

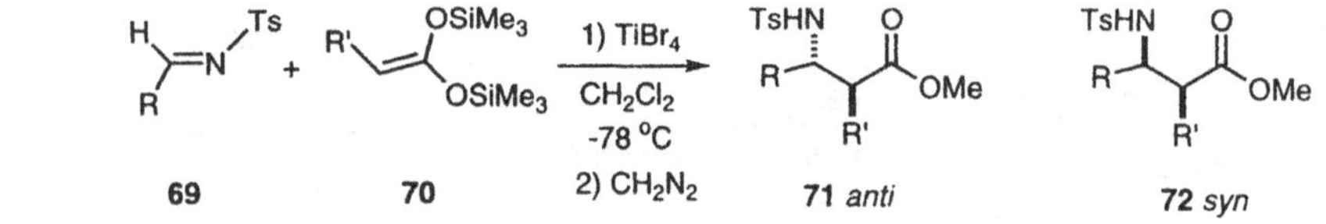

69 70 71 *anti* 72 *syn*

R	R'	Yield	*anti*:*syn*
R=Ph	R'=Me	88%	92:8
*p*MeOPh		70%	90:10
*p*NO$_2$Ph		85%	91:9
2-furyl		91%	88:12
E-CH=CHPh		22%	87:13
Ph	Et	85%	87:13
Ph	iPr	85%	81:19
Ph	Ph	53%	75:25

Scheme 15

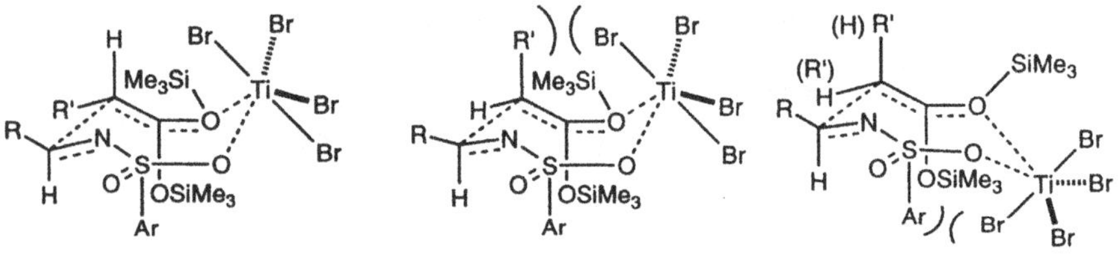

73 74 75

Fig. 1

destabilized relative to **73**, which leads to *anti* compounds **71**, due to a bad steric interaction between a bromine ligand on titanium and the ketene acetal substituent R′. The alternative transition states **75** are both less stable than **73** for similar steric reasons.

This supposition raises an interesting point with regard to the site of Lewis acid complexation of *N*-sulfonyl imines. Weinreb and Sisko [38] have observed that ^{1}H and ^{13}C NMR spectra of alkyl *N*-tosyl imines in the presence and absence of a Lewis acid are virtually identical. It would be anticipated that if Lewis acid complexation in fact occurred at nitrogen, one should observe a significant downfield shift [39] of the imino proton or carbon. Therefore, it may be that in general Lewis acids prefer to coordinate at the sulfonyl group of this type of imine, although additional work is necessary to verify the exact position of Lewis acid complexation.

A few additional scattered examples have been reported of additions of enol derivatives to *N*-sulfonyl imines and iminium ions. For instance, Kobayashi et al. [40] have found that the condensation shown in Eq. (20) is catalyzed effectively by a lanthanide triflate in excellent yield.

The α-methoxy sulfonamides **76** [Eq. (21)] react with the silyl ketene acetal **77** using trimethylsilyl triflate as catalyst to afford adducts **78** in good yields [29]. $TiCl_4$ and $SnCl_4$ were found to be ineffective catalysts in this transformation. With the α-hydroxy sulfonamide corresponding to **76**, only complex mixtures were obtained. Similarly, the α-methoxy sulfonamide **79** condensed with the silylenol ether of acetophenone to give **80** [34] [Eq. (22)]. The best catalysts for this reaction were $SnCl_4$, $TiCl_4$ and $FeCl_3$. ZnI_2 and BF_3 etherate gave lower yields of ketone **80**. The acetate **79** was also useful in this reaction when $SnCl_4$ was used.

An interesting and highly stereoselective reaction of dimethoxy cyclopropane derivative **81** with some aromatic *N*-tosyl imines was recently described by Saigo and coworkers [41] (Scheme 16). In the presence of $TiCl_4$, compound **81** condenses with *N*-sulfonyl imines to stereoselectively produce lactams **84** and **85**, with the *cis* isomer being the predominant product. It is likely that the dimethoxy cyclopropane initially opens to zwitterionic ester enolate **82**, which adds to the imine to yield intermediate **83**. The rationale presented for the stereoselectivity in condensation of enolate **82** with the imines is similar to that described for the reactions in Schemes 14 and 15, cf. Fig. (1).

One additional example of a reaction of an enol-type derivative with an in situ-produced *N*-sulfonyl iminium species involves the $TiCl_4$-promoted reaction of **22** with acetoxy furan **86** to yield butenolide **87** [25b] [Eq. (23)].

Scheme 16

(23)

4.4
Alkenes

Although there is extensive literature on reactions of *N*-acyl imines with simple olefins [1b, c], surprisingly little related chemistry of *N*-sulfonyl imines is available. In some studies of intramolecular cyclizations of *N*-sulfonyl imines with alkenes, Weinreb and coworkers [11] first demonstrated the feasibility of such a process. In these studies, the Kresze strategy was utilized for generation of the requisite *N*-sulfonyl imines [6, 7] [cf Eq. (6)]. Therefore, aldehyde olefin **88** was treated with *N*-sulfinyl-*p*-toluenesulfonamide in the presence of a Lewis acid (Scheme 17), presumably first forming an imine complex **89**. Subsequent cyclization of **89** would then afford cation **90**. In the presence of an equivalent of BF₃ etherate, fluoride is transferred to afford the halogenated product **91** as a single stereoisomer. When less Lewis acid was used, a mixture of fluoride **91** and olefin **92** was formed. If ferric chloride was used as catalyst, chloro sulfonamide **93** was produced as a single stereoisomer.

Cyclization of aldehyde olefin **94** was also investigated, and was found to lead to a 1:1 mixture of bridged products **99** and **100** (Scheme 18). This transformation probably occurs through sulfonyl imine Lewis acid complexes cyclizing via conformations **95** and **96** to carbonium ions **98** and **97**, respectively. Proton elimination from these intermediates would afford the observed epimeric sulfonamide alkenes.

The aldehyde alkene substrate **101** [Eq. (24)] was found to cyclize to an epimeric mixture of *cis*-decalin derivatives **102** and **103**. Similarly, acyclic substrate

Scheme 17

Scheme 18

(24)

104 [Eq. (25)] produced a mixture of *trans* and *cis* products **105** and **106**. It appears that halogenated products are formed in olefin cyclizations only in those systems where proton elimination from the intermediate carbonium ion is relatively slow.

An unsuccessful attempt was made to apply this methodology to a total synthesis of the *Ergot* alkaloid lysergic acid (Scheme 19) [42]. Therefore, aldehyde olefin **107** was prepared and was exposed to the Kresze conditions with the

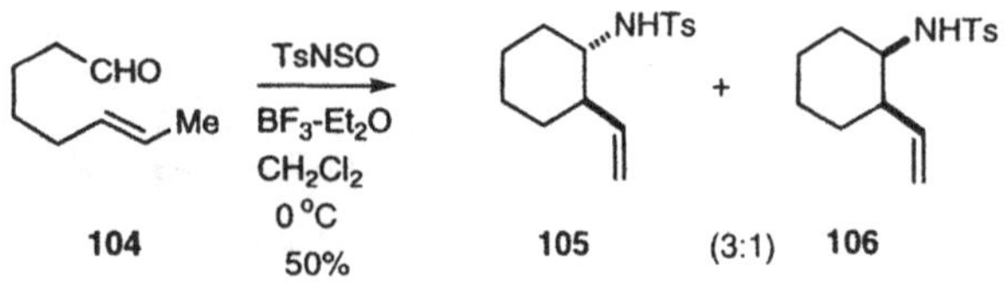

$$(25)$$

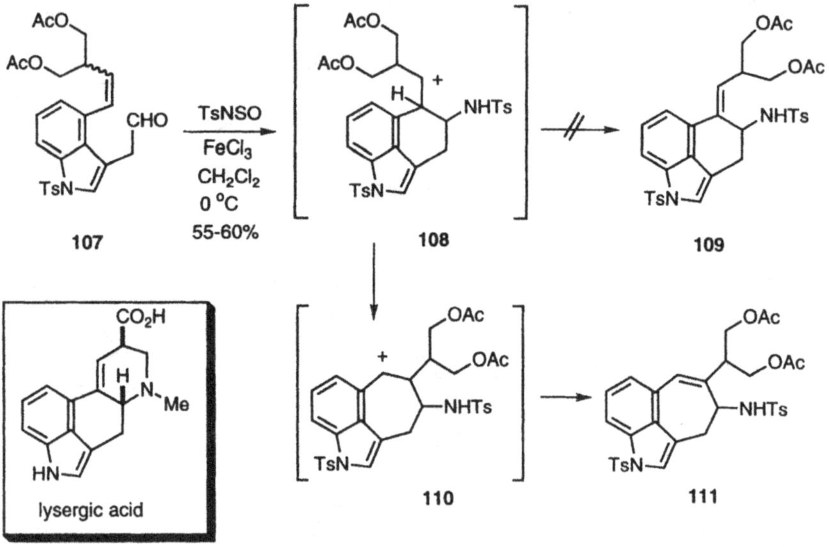

Scheme 19

intent of generating the required tricyclic sulfonamide **109** via carbonium ion **108**. However, it appears that **108** is prone to rearrangement to the carbonium ion **110**, since the undesired ring expanded system **111** is the actual product isolated from the cyclization. The structure of **111** was confirmed by X-ray crystallography.

Recently Ahman and Somfai [43] have used an *N*-sulfonyl iminium ion-alkene cyclization as a key step in an enantioselective total synthesis of the alkaloid anatoxin A (Scheme 20). α-Hydroxy sulfonamide **55** was prepared from L-pyroglutamic acid (cf Scheme 12) and was transformed in 6 steps into enone **112**. Exposure of **112** to acid led to a mixture of bridged enone **114** and β-chloro ketone **113**. The latter compound could be converted into the desired enone with DBU. Detosylation of **114** provided the natural product (+)-anatoxin A.

4.5
Allyl Silanes

A number of examples of inter- and intramolecular additions of allyl silanes to *N*-sulfonyl imines have been reported. Weinreb and coworkers have combined the Kresze methodology for forming *N*-sulfonyl imines and subsequent additi-

Scheme 20

ons of allyl silanes [15] into a "one pot" procedure. In preliminary studies, it was found that propionaldehyde can be converted to the N-tosyl imine using N-sul-finyl-p-toluene sulfonamide and a Lewis acid, followed by addition of allyl tri-methylsilane to afford the allyl sulfonamide (Scheme 21). For this particular reaction, $SnCl_4$ proved to be the best Lewis acid. Additional examples of this transformation are given in Table 1. In general, $SnCl_4$ and $FeCl_3$ hexahydrate pro-ved to be the best catalysts. Other Lewis acids, such as $AlCl_3$, $TiCl_4$ and $ZnCl_2$, gave poor yields of allylation products.

Lewis Acid	Solvent	Temperature	Isolated Yield (%)
BF_3-Et_2O	PhH	0 °C to rt	60
BF_3-Et_2O	CH_2Cl_2	-30 °C	25
BF_3-Et_2O	PhMe/CH_2Cl_2	-30 °C	42
$SnCl_4$	PhH	0 °C to rt	95
$FeCl_3$-$6H_2O$	PhMe	0 °C to rt	61

Scheme 21

A few other studies of allylations of N-sulfonyl imines by allylic silanes have been published. Shono et al. found that azetidine derivative **22** can be converted to allyl compound **115** [25b] [Eq. (26)]. The functionalized sulfonamides **116** [Eq. (27)] have all been alkylated with allyl trimethylsilane and Lewis acids [34]. In general, the best yields of **117** were obtained with $SnCl_4$, BF_3 etherate, ZnI_2 and $FeCl_3$ as catalysts. Lower yields were obtained with $TiCl_4$ and Et_2AlCl. Simi-larly, the α-hydroxy and α-methoxy sulfonamides **51** could be alkylated to give

Table 1 Reactions of *in situ* Generated *N*-Tosyl Imines with Allyl Silanes

Aldehyde	Allyl Silane	Conditions	Product(s)	% Yield
EtCHO	Me₂C=CH–CH₂SiMe₃	FeCl₃·6H₂O PhMe/0 °C	Et–CH(NHTs)–C(Me)₂–CH=CH₂	71
EtCHO	cyclopentenyl-SiMe₃	FeCl₃·6H₂O PhMe/0 °C-rt	Et–CH(NHTs)–(cyclopentenyl) (2:1)	72
iPrCHO	allyl-SiMe₃	FeCl₃·6H₂O PhMe/0 °C	iPr–CH(NHTs)–CH₂CH=CH₂	93
iPrCHO	Ph–CH=CH–CH₂SiMe₃	SnCl₄/PhH/0 °C-rt	iPr–CH(NHTs)–CH(Ph)–CH=CH₂	69
Me(CH₂)₅CHO	allyl-SiMe₃	FeCl₃·6H₂O PhMe/0 °C	Me(CH₂)₅–CH(NHTs)–CH₂CH=CH₂	92
PhCHO	allyl-SiMe₃	SnCl₄ PhMe/rt	Ph–CH(NHTs)–CH₂CH=CH₂	89

$$\textbf{22} \xrightarrow[\substack{-70\,°C \\ 62\%}]{\substack{\text{allyl-SiMe}_3 \\ \text{TiCl}_4 \\ \text{CH}_2\text{Cl}_2}} \textbf{115} \qquad (26)$$

$$\textbf{116}\ (R=\text{OH, OMe, OAc, SPr}) \xrightarrow[\substack{\sim-60\,°C}]{\substack{\text{allyl-SiMe}_3 \\ \text{Lewis acid} \\ \text{CH}_2\text{Cl}_2}} \textbf{117} \qquad (27)$$

116 R=OH, OMe, OAc, SPr   **117**

119 in high yields using TiCl₂(O*i*Pr)₂, SnCl₄, TiCl₄, FeCl₃, BF₃ etherate and CF₃CO₂H as catalysts [29] (Scheme 22). Ti(O*i*Pr)₄ was not a strong enough acid to promote the reaction and PPTS led only to enamides **118**.

Lu and Zhou [44] have utilized a reaction between an *N*-sulfonyl iminium ion and allyl trimethylsilane in enantioselective total syntheses of two piperidine alkaloids (Scheme 23). The initial step in this approach involved a modified Sharpless kinetic resolution of furfuryl sulfonamide **120**, leading to *R*-amide **121**

Scheme 22

Scheme 23

and *S*-oxidation product **122** [45]. The piperidone derivative **122** could be converted in high yield to the *N*-sulfonyl iminium precursor **123**. Allylation of **123** showed reasonable *cis*-stereoselectivity giving a 16:84 mixture of **124:125**. It was then possible to convert the major isomer **125** into (2*S*, 6*R*)-dihydropinidine and also into (+)-azimic acid.

Somfai and Ahman have applied an intramolecular allyl silane addition to an *N*-sulfonyl iminium ion as a key step in an alternative synthesis of (+)-anatoxin A [43]. Thus, L-pyroglutamic acid-derived compound **55** was homologated to aldehyde **126** and then to allyl silane **127** (Scheme 24). Using titanium tetrachloride, **127** could be cyclized in good yield to bicyclic olefin **128**, which was converted to the alkaloid.

Scheme 24

An intramolecular allyl silane/*N*-sulfonyl iminium ion cyclization has also been used as a pivotal step in an approach to the tricyclic core of the unique marine alkaloid sarain A [46]. The starting material was aziridine ester **129** (Scheme 25) which was elaborated to amide **130**. An important step in the synthetic strategy was thermolysis of **130** to an azomethine ylide, which underwent stereospecific intramolecular 1,3-dipolar cycloaddition with the *Z*-alkene to produce bicyclic lactam **131** [47]. This compound was then elaborated into allyl silane **132**. It was then possible to replace the lactam *N*-benzyl functionality with a tosyl moiety, leading to **133**, and subsequent reduction of the carbonyl group afforded the desired cyclization precursor *α*-hydroxy sulfonamide **134**. Exposure of **134** to ferric chloride promoted cyclization to a single stereoisomeric tricyclic amino alkene **136** having the requisite sarain A nucleus. It is believed that the intermediate *N*-sulfonyl iminium ion cyclizes via the conformation shown in **135**.

4.6
Vinyl Silanes

Despite a considerable amount of recent work on reactions of vinyl silanes with various kinds of imines [48, 49], scant attention has been paid to *N*-sulfonyl imines in this area. A single study of a vinyl silane/*N*-sulfonyl imine reaction has been published by McIntosh and Weinreb in the context of an approach to the total synthesis of [1, 3]-dioxolophenanthrene structural types of *Amaryllidaceae* alkaloids such as narciclasine (**137**), lycoricidine (**138**) and pancratistatin (**139**) [50]. The substrate used in this approach was vinyl silane aldehyde **140**, prepared enantiomerically pure in a straightforward manner from L-arabinose (Scheme 26). The *N*-tosyl imine derived from this aldehyde could be generated in two different ways. The first involved combination of **140** with *N*-sulfinyl-*p*-toluenesulfonamide at 80 °C, followed by exposure of the imine to BF$_3$ etherate at 0 °C, leading to a single cyclization product **142** in 36 % yield. The second procedure was to simply react aldehyde **140** with *p*-toluenesulfonamide and BF$_3$ etherate (–78 °C -rt) to afford a 9.5:1 mixture of **142:144** in ~80 % yield. It was pro-

sarain A

129

130

1) *o*-DCB
320 °C
2) TsOH
MeOH
70%

131 OH

4 steps

132 SiMe₃

1) Na/NH₃
*t*BuOH
95%
2) LHMDS
TsCl, DMAP
71%

133 SiMe₃

DIBALH
93%

134 SiMe₃

FeCl₃/CH₂Cl₂
-78 °C to rt

135

61%

136

Scheme 25

posed that these cyclizations occur via a Lewis acid-complexed *N*-tosyl aldimine, and that conformation **141**, leading to **142**, is favored over the conformation **143**, which affords product **144**, in order to minimize developing gauche interactions.

A variation of this cyclization as shown in Eq. (28) was also effected. Treatment of aldehyde **140** with *N*-methyl-*p*-toluenesulfonamide and BF₃ etherate led

Scheme 26

(28)

to **146** as a single stereoisomer in good yield, probably via the *N*-sulfonyl iminium ion **145**.

The cyclization product **142** has the desired absolute configuration for the
alkaloids **137–139** and was transformed as described in Scheme 27. Thus,
condensation of **142** with acid chloride **147** gave *N*-tosyl amide **148**. This compound underwent intramolecular Heck cyclization to yield tetracycle **149** which
is a derivative of (+)-lycoricidine (**138**).

4.7
Hydrides

Some limited work has been reported by the Weinreb group on hydride reductions of *N*-sulfonyl imines. The ketone group of bridged lactone **150**, which was

Scheme 27

a key intermediate in total syntheses of the antitumor antibiotics actinobolin and bactobolin [51, 52], was sufficiently reactive that it could be converted to the SES and PMS sulfonyl imines **151** using the Kresze methodology (Scheme 28). Sodium cyanoborohydride reduction of these imines was stereoselective and provided sulfonamide lactones **152**.

$$R = PMS = pMeOPhCH_2SO_2-$$
$$R = SES = Me_3SiCH_2CH_2SO_2-$$

Scheme 28

Weinreb and coworkers have also developed a simple "one-pot" procedure for reductive sulfonamidation of both aromatic and aliphatic aldehydes [14]. Again using the Kresze methodology an aldehyde could be converted to a Lewis acid-complexed *N*-tosyl imine which in the presence of triethylsilane was reduced to a sulfonamide in good yields [Eq. (29)].

$$\text{R = alkyl, aryl} \tag{29}$$

4.8
Organometallics

Additions of basic organometallic reagents, such as Grignards and organo lithiums, to imines is often troublesome due to competing deprotonation and electron transfer processes [53]. However, due to their high electrophilicity, N-sulfonyl imines have proven to be excellent partners in organometallic additions. Sisko and Weinreb [13] have developed a "one-pot" procedure for this transformation, starting from aryl and aliphatic aldehydes, using Kresze methodology. With aliphatic aldehydes it is possible to generate an N-sulfonyl imine in situ using an N-sulfinyl sulfonamide at room temperature with no added catalyst. Addition of a wide variety of Grignards and lithium reagents to these imines gives good yields of adducts [Eq. (30)]. With aromatic aldehydes, the reaction with sulfinyl sulfonamides is too slow to be useful unless a Lewis acid is used, and therefore a modified procedure is necessary [Eq. (31)].

$$alkyl\text{---}CHO \xrightarrow[\text{1-2 h/rt}]{\substack{\text{TsNSO} \\ \text{CH}_2\text{Cl}_2}} \left[alkyl\diagup\!\!\!\diagdown^{NTs} \right] \xrightarrow[\text{RLi}]{\substack{\text{RMgX} \\ \text{or}}} alkyl\diagdown\!\!\!\!\diagup\substack{NHTs \\ R} \qquad (30)$$

$$Ar\text{---}CHO \xrightarrow[\substack{\text{CH}_2\text{Cl}_2 \\ \text{5-6 h/0 °C}}]{\substack{\text{TsNSO} \\ \text{BF}_3\text{-Et}_2\text{O}}} \left[Ar\diagup\!\!\!\diagdown^{\substack{\text{BF}_3 \\ \cdot \\ NTs}} \right] \xrightarrow[\text{RLi}]{\substack{\text{RMgX} \\ \text{or}}} Ar\diagdown\!\!\!\!\diagup\substack{NHTs \\ R} \qquad (31)$$

Reetz and coworkers have used this methodology in a stereoselective synthesis of vicinal diamines [Eq. (32)] [54]. Enantiomerically pure α-amino aldehydes 153, which are available from α-amino acids, can be converted into N-tosyl imines 154 using the N-sulfinyl sulfonamide procedure [13]. Addition of a wide range of Grignard reagents to imines 154 gave >90:10 mixtures of adducts 155:156 in very good yields. One rationale for formation of *erythro* isomers 155 as the major products would be to invoke a Felkin-Anh model for the addition. A complimentary process which was also described involved addition of organolithium reagents in the presence of a lanthanide salt to the N-benzyl imines from aldehydes 153, which afforded mainly the *threo* stereoisomers 156 as the primary products.

$$\underset{\substack{\textbf{153 } R=Me, Bn, \\ iPr, iBu}}{\overset{Bn_2N}{\underset{R}{\diagup}}\!\!\diagdown_{H}^{O}} \xrightarrow[\text{2-3 h/rt}]{\substack{\text{TsNSO} \\ \text{CH}_2\text{Cl}_2}} \left[\underset{\textbf{154}}{\overset{Bn_2N}{\underset{R}{\diagup}}\!\!\diagdown_{H}^{NTs}} \right] \xrightarrow[\text{70-95%}]{\substack{R'MgX}} \underset{\substack{\textbf{155} \\ R'=alkyl, aryl, \\ allyl, vinyl}}{\overset{Bn_2N}{\underset{R}{\diagup}}\!\!\diagdown_{R'}^{NHTs}} + \underset{\textbf{156}}{\overset{Bn_2N}{\underset{R}{\diagup}}\!\!\diagdown_{R'}^{NHTs}} \qquad (>90:<10) \qquad (32)$$

4.9
Reaction with β-Hydroxy Aldehydes

N-Sulfonyl imines undergo an interesting and unprecedented reaction with various β-hydroxy aldehydes [55]. Thus, treatment of an *N*-sulfonyl imine **158**, produced by the sulfinyl sulfonamide method with a β-hydroxy aldehyde **157** in the presence of BF$_3$ etherate, afforded *trans*–2,6-disubstituted 3,6-dihydro-2*H*–1,3-oxazines **160** (Scheme 29). It is believed that the reaction of **157** with imine **158** initially affords an adduct **159**, which subsequently undergoes cyclo-dehydration to the observed products. Although one would expect that intermediate **159** is a complex mixture of stereoisomers, the fact that only the *trans* 2,6-disubstituted heterocycle is isolated may indicate that some type of acid-promoted amido acetal equilibration may be taking place to produce the thermodynamically most stable product. ^{1}H NMR NOE studies and X-ray crystallography indicate that these dihydrooxazines have the conformation shown. Interestingly, the aryl group of the sulfonamide is in a quasi axial position and blocks one face of the molecule, directing the stereochemistry of some of the reactions of this system.

R=*n*Pr, R'=Et, R"=*n*Pr   83%
R=Et, R'=Et, R"=*n*Pr   75%
R=Me, R'=Et, R"=*n*Pr   73%
R=*n*Pr, R'=Me, R"=Et   87%
R=Me, R'=Me, R"=Et   71%
R=*n*Pr, R'=H, R"=Me   40%
R=Et, R'=H, R"=Me   51%
R=*n*Pr, R'=H, R"=*i*Pr   47%
R=*n*Pr, R'=H, R"=*n*C$_{14}$H$_{29}$ 42%
R=Et, R'=Me, R"=Me   82%

Scheme 29

Dihydrooxazines of this type were essentially unknown previously, and studies were therefore undertaken to explore their potential as synthons in the stereoselective synthesis of 1,3-amino alcohols. For example, oxidations of the ring double bond were investigated. Thus, hydroxylation of **161** with osmium tetraoxide was stereoselective, affording diol **162** (Scheme 30) resulting from attack on the face of the double bound *anti* to the aryl sulfonyl group (cf. **160**). This diol could be converted to an epimeric mixture of hydroxy nitriles **163** and **164** via an *N*-sulfonyl iminium intermediate.

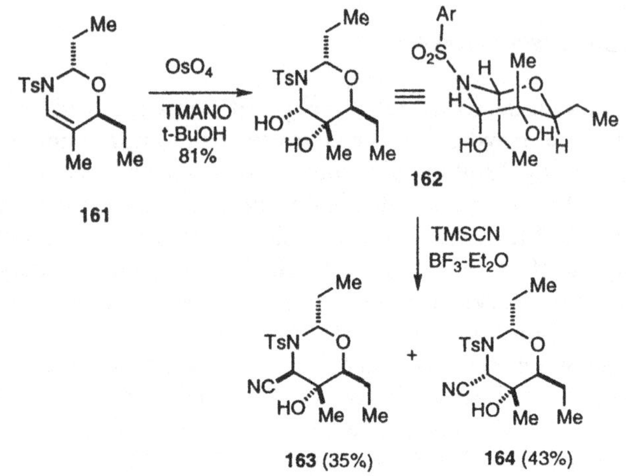

Scheme 30

On the other hand, oxidation of **161** with dimethyl dioxirane gave a 1:9 mixture of diols **165** and **166** (Scheme 31). Surprisingly and inexplicably the major product **166** results from attack on the double bond face *syn* to the arylsulfonyl moiety. Diol **166** could be converted to monoacetate **167** which underwent stereoselective alkylation with allyl trimethylsilane to yield **168**. Similarly, acetate **167** could be converted to a single nitrile **169**. Both of these transformations involve axial attack *anti* to the arylsulfonyl group on an intermediate *N*-sulfonium iminium ion.

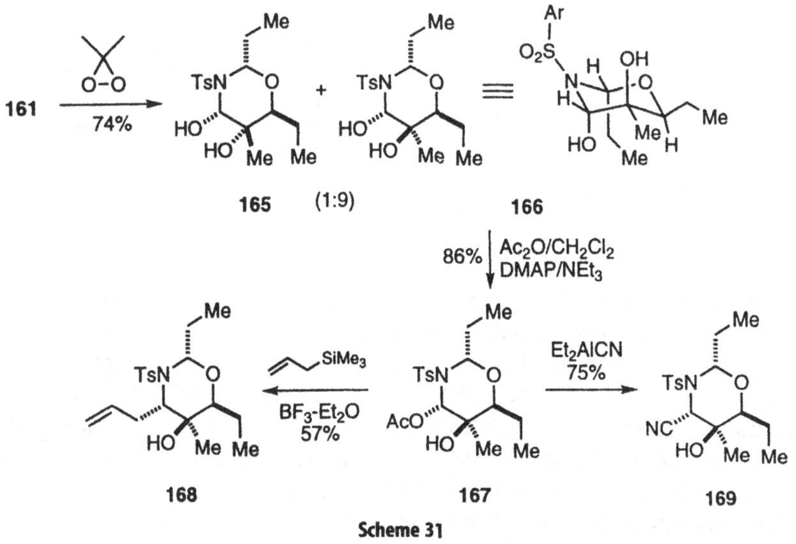

Scheme 31

4.10
Aromatic and Heteroaromatic Amidoalkylations

Although the literature abounds with examples of aromatic amidoalkylations with *N*-acyl imines [1a, 56], virtually nothing has been done in this area with *N*-sulfonyl imines. An amidoalkylation of this type involves reaction of α-methoxy sulfonamide **170** with anisole to afford alkylation products **171** and **172** [Eq. (33)] [25c]. In a heteroaromatic version of this process, thiophene was

found to react thermally with imine **173** to afford adduct **174** [Eq. (34)] [57]. The tosyl imine from methyl glyoxylate reacts readily with furans to give amidoalkylation products [Eq. (35)] [58]. Many years ago [59] it was reported that α-picoline reacts with sulfanilamide and paraformaldehyde to produce adduct **175** [Eq. (36)]. It is possible that an *N*-sulfonyl imine is an intermediate here.

5
[4+2]-Cycloadditions

5.1
Heterodienophiles

5.1.1
N-Sulfonyl Imines of Chloral and Fluoral

The earliest reports of imines acting as dienophiles in Diels-Alder reactions involved *N*-sulfonyl imines prepared from chloral and fluoral using the orginal Kresze [6] sulfinyl sulfonamide procedure. Thus, the *N*-tosyl imine **176** from chloral reacts under mild conditions with a variety of 1,3-dienes to give cyclo-adducts [60–62]. These reactions have been found to show excellent regioselectivity. For example, combination of **176** with *E*-piperylene gives only adduct **177**, whereas 2-methoxybutadiene leads exclusively to **178** (Scheme 32) [61, 62]. This selectivity has been rationalized [1 d, 62, 64] by assuming that these dienophiles are highly polarized, and that this polarization is reflected in the transition state for cycloaddition.

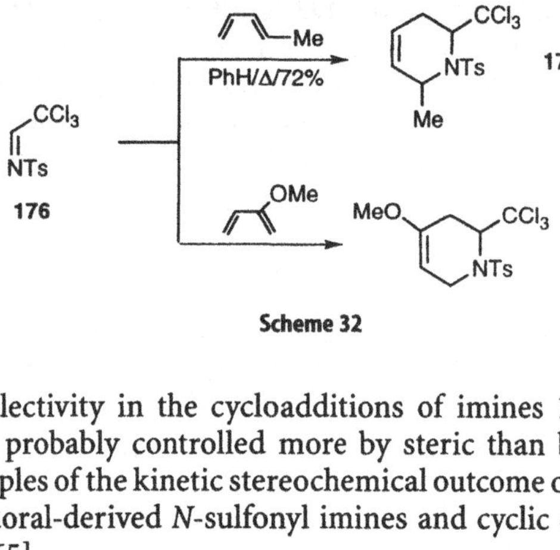

Scheme 32

The stereoselectivity in the cycloadditions of imines **176** and **179** is only modest, and is probably controlled more by steric than by electronic effects [63 b, 65]. Examples of the kinetic stereochemical outcome of cycloaddition with chloral- and fluoral-derived *N*-sulfonyl imines and cyclic dienes are shown in Eq. (37) – (39) [65].

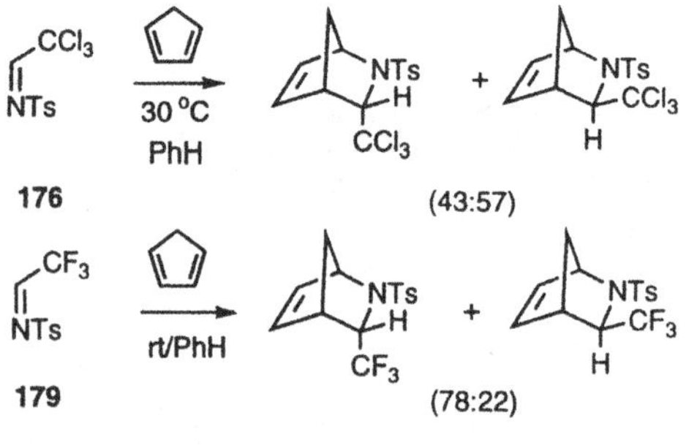

$$ \text{(39)} $$

(57:43)

5.1.2
N-Sulfonyl Imines of Glyoxylates

The vast majority of examples of hetero Diels-Alder reactions of *N*-sulfonyl imines involve glyoxylate-derived compounds. In general, these glyoxylate *N*-sulfonyl imines have been made using the Kresze protocol [6, 7], although recently Holmes and coworkers have found that commercially available tosyl isocyanate can be used in reaction with methyl glyoxylate in place of the *N*-sulfinyl sulfonamide [58]. As with the chloral derivatives, *N*-sulfonyl glyoxylates show excellent regioselectivity in Diels-Alder reactions with unsymmetrical dienes. Two examples of this selectivity are shown in Eq. (40) [7] and (41) [58, 66], and can once again be conveniently rationalized by assuming the cycloadditions proceed via polarized transition states [1 d].

$$ \text{(40)} $$

$$ \text{(41)} $$

Although the Diels-Alder reactions of glyoxylate *N*-sulfonyl imines often show high stereoselectivity, the results are inconsistent and difficult to rationalize at this point. In the case of cyclopentadiene [Eq. (42)] [58, 68] and 1,3-cyclohexadiene [Eq. (43)] [68] the kinetic products of cycloaddition are the carboxylate *exo* isomers. With the acyclic diene 1,3-dimethylbutadiene, a mixture of *exo* and *endo* products is observed with *exo* predominating [Eq. (44)] [58].

In contrast to the above results, the acyclic dienes in Eqs. (45) [69] and (46) [70] afford only the products of *endo* carboxylate addition. It is difficult to develop a coherent model to explain these diverse stereochemical results. In addition, it should be noted that rapid geometrical inversion of the *N*-sulfonyl imines makes the reacting configuration of these species an unknown which complicates analysis [30, 31].

Hamada and coworkers [70] have examined the facial selectivity of glyoxylate *N*-sulfonyl imine cycloadditions with acyclic dienes bearing a stereogenic cen-

$$(42)$$

$$(43)$$

$$(44) \quad (2:1)$$

$$(45)$$

$$(46)$$

180   **181**   **182**

ter at an allylic position [Eq. (46)]. Thus, cyclization with diene **180** afforded a single product **182** in good yield. These results suggest that the reaction proceeds via the diene rotamer as shown in **181** [71]. Interestingly, the selectivity in this type of imino Diels-Alder reaction is much higher than with various all carbon and azo dienophiles [71].

The same group has applied this imino Diels-Alder reaction in an enantioselective total synthesis of the alkaloid (–)-cannabisativine (Scheme 33) [70b]. An initial Sharpless epoxidation of allylic alcohol **183** provided enantiomerically pure compound **184**, which could be converted in four steps to alcohol **185**. This compound could then be relayed into the requisite diene **186**. Imino Diels-Alder reaction of **186** led to a single cycloadduct **188**, presumably via a transition state like **187**. It was then possible to homologate the ester functionality of **188** via the corresponding aldehyde to acetal **189**. This intermediate could be converted in several steps into **190**. Another key step in the strategy was epimerization via a retro Michael reaction leading to aldehyde **191**, which could be transformed in five steps into (–)-cannabisativine.

Scheme 33

Another nice application of this glyoxylate *N*-sulfonyl imine type of Diels-Alder methodology in the realm of alkaloid total synthesis has been described by Holmes and coworkers [72]. An approach to the piperidine alkaloid (+)-isoprosopinine B is outlined in Scheme 34. It was found that cycloaddition of the glyoxylate imine with siloxy diene **192**, followed by acidic hydrolysis, produced a mixture of *endo* adduct **193** and the *exo* product **194** with the latter predominating. The major *exo* product **194** underwent regioselective Baeyer-Villiger oxidation to afford lactone **195** along with a trace of the regioisomeric compound. Hydride reduction of the lactone and ester groups of **195** led to triol **196**, which, by a series of transformations, led to (+)-isoprosopinine B.

Two groups have investigated *N*-sulfonyl imines of glyoxylate esters, derived from scalemic alcohols, in Diels-Alder reactions [67b, 73]. Prato and coworkers reported that glyoxylate *N*-sulfonyl imines bearing (–)-menthyl, (–)-bornyl and (–)-8-phenylmethyl auxiliaries reacted with cyclopentadiene either thermally or using Lewis acids to give only very modest diastereomeric product ratios (56:44, 53:47, 60:40, respectively) [67b].

193 (18.5%)
endo

194 (47%)
exo

CH$_3$CO$_3$H

47%

195

LiAlH$_4$

80%

196

7 steps

isoprosopinine B

Scheme 34

A more successful approach to diastereofacial selectivity in this type of cycloaddition was described by Holmes, et al. [73]. It was discovered that both the (S)-lactate-derived imine **197** and the (R)-pantothenate **198** showed useful levels of diastereoselectivity in cycloadditions with cyclopentadiene (Scheme 35). Although thermal reactions of imines **197** and **198** with cyclopentadiene showed relatively low facial diastereoselectivity, Lewis acid catalysis improved the selectivity significantly. Diethylaluminium chloride proved to be the best catalyst, providing the product ratios shown in Scheme 35. Other Lewis acids, such as TiCl$_4$, SnCl$_4$ and BF$_3$ etherate, caused decomposition, whereas Al(OEt)$_3$, MgBr$_2$, ZnCl$_2$, Me$_2$AlCl and iBu$_2$AlCl gave lower selectivities. The model used to rationalize these results is shown for the (S)-lactate case in Fig. 2. The Lewis acid-complexed ester is believed to exist in the *s-trans* conformation, and attack of the diene occurs from the less hindered *si* face [74].

197

0.1 eq
Et$_2$AlCl

0.3 eq
Et$_2$AlCl

PhMe
-78 °C
50-60%

198

199

200

R=(S)-lactate 12:88

R=(R)-pantothenate 85:15

Scheme 35

Fig. 2

Whiting and coworkers have taken a somewhat different approach to effecting diastereoselective glyoxylate *N*-sulfonyl imine Diels-Alder reactions [75]. This group has incorporated a chiral auxiliary into the sulfonyl moiety using camphor sulfonic acid as starting material. The requisite imine **203** was prepared from sulfonamide **201** via bromination to **202** and HBr elimination (Scheme 36). Diels-Alder cycloadditions were conducted using Danishefsky diene **204** leading to enones **205a** and **205b** after acidic workup (Eq. 47). Modest diastereoselectivity was achieved in the absence of a Lewis acid catalyst, with the best result being a 2.04:13 ratio of diastereomers **205a:205b** in CCl_4 at −15°C. With catalytic amounts of various Lewis acids at −75°C in toluene, ratios of diastereomers **205a:205b** ranged from 2.30:1 with titanium tetraisopropoxide to a reversed ratio of 1:1.44 with diethylaluminium chloride. No mechanistic model was offered to rationalize these results.

201   **202**   **203**

Scheme 36

$$204 \quad + \quad 203 \quad \xrightarrow{\text{Lewis acid}} \quad 205 \tag{47}$$

204   **203**   **205a** *=*R*
                              205b *=*S*

5.1.3
Other N-Sulfonyl Imines

Two examples of [4+2]-cycloadditions of cyclic *N*-sulfonyl imines have been described [76]. It was found that imine **206** reacts with Danishefsky diene **204** to produce cycloadduct **207** [Eq. (48)]. Similarly, chloro-substituted imine **208** was reported to add to diene **204** leading ultimately to dienone **209** [Eq. (49)].

Weinreb and Sisko have reported the first examples of Diels-Alder reactions of N-tosyl imines derived from enolizable aldehydes [12]. The imines were generated in situ from the aldehyde, N-sulfinyl-p-toluenesulfonamide and boron trifluoride etherate. Two examples of these cycloadditions are shown in Egs. (50) and (51). It was also possible to effect the cycloaddition intramolecularly [Eq. (52)].In a recent report [77], Furukawa et al. found a novel method for generation of N-tosyl imines which have been trapped as [4+2]-cycloadducts. 1,8-Naphthalene dithiol can be converted in two steps into sulfilimines **210** [Eq. (53)]. This heterocycle rearranges in the presence of BF_3 etherate into **211**, which then leads to N-tosyl imine **212**. This species can subsequently be used in Diels-Alder reactions to produce cycloadducts in moderate yields.

210 R=Me, Et, hexyl, Ph   **211**   **212**   (53)

5.2
Heterodienes

In a recent study, Boger and coworkers have thoroughly probed cycloadditions of *N*-sulfonyl α, β-unsaturated imines with electron-rich olefins [78]. The *N*-sulfonyl imines were prepared in most cases from the aldoximes or ketoximes by the Hudson methodology [18]. Alternatively, some imines could be synthesized directly from α, β-unsaturated aldehydes and the sulfonamide using a Lewis acid catalyst. It was found that cycloadditions with these *N*-sulfonyl heterodienes could generally be effected at room temperature or below at high pressure or could be effected thermally. For example, azadiene **214** reacts with ethyl vinyl ether to give adduct **213** (Scheme 37). This reaction, as all others in this series, was found to be totally regioselective and >95% *endo* selective. Similarly, diene **214** reacts with methoxy allene to give *endo* adduct **215**. It was also found that aldimines react faster than ketimines in these cycloadditions [Egs. (54) and (55)].

213   **214**   **215**

Scheme 37

(54)

(55)

Interestingly, it was shown that substitution of the 1-azadiene at the 2, 3 or 4 positions by an electron-withdrawing group led to an acceleration of the rate of cycloaddition. The 3-substituted diene **216** reacted very rapidly with 1,1-dimethoxyethylene to afford adduct **217** [Eq. (56)]. Similarly, 2-substituted diene **219** undergoes highly *endo*-selective cycloadditions. Thus, reaction of **219** with ethyl vinyl ether gave **220**, whereas with *E*-1-ethoxy propene adduct **218** was formed (Scheme 38). Reactions of 4-substituted-1-azadienes also proved to be rapid and *endo*-selective. For example, reaction of diene **223** with ethyl vinyl ether led to *endo* cycloadduct **224** (Scheme 39). The cycloaddition of **223** with styrene was less stereoselective giving mixtures of *endo* product **221** and *exo* adduct **222** in varying ratios depending upon conditions.

These results, including the fact that the cycloaddition rates are independent of solvent polarity, are consistent with a concerted LUMO-diene controlled process. This assumption is also supported by calculations [79]. Interestingly, the

$$(56)$$

216   **217**

Scheme 38

218   **219**   **220**

221   **222**   **223**   **224**

CH$_2$Cl$_2$/21 °C   6.5:1 (45%)
PhH/80 °C   11:1 (48%)

Scheme 39

high *endo*-alkoxy selectivity can be ascribed to an anomeric-like effect in the transition state.

This methodology has been elegantly utilized as a key step in a total synthesis of the antitumor antibiotic fredericamycin (Scheme 40) [80]. Thus, 1-azadiene **225** was found to react with olefin **226** to yield adduct **227** as a 1:1 mixture of stereoisomers. This compound could then be aromatized to pyridine **228**. In an interesting transformation, 4-methylpyridine **228** reacts with cyclopentenone via an initial Michael addition, followed by a Claisen condensation, to afford tricycle **229**. This compound could be aromatized and *O*-benzylated to produce ketone **230**, which was homologated to nitrile ester **231**. The ester functionality of **231** could be transformed to *E,E*-diene **232**. It was then possible to utilize this DEF fragment in a sequence leading to fredericamycin.

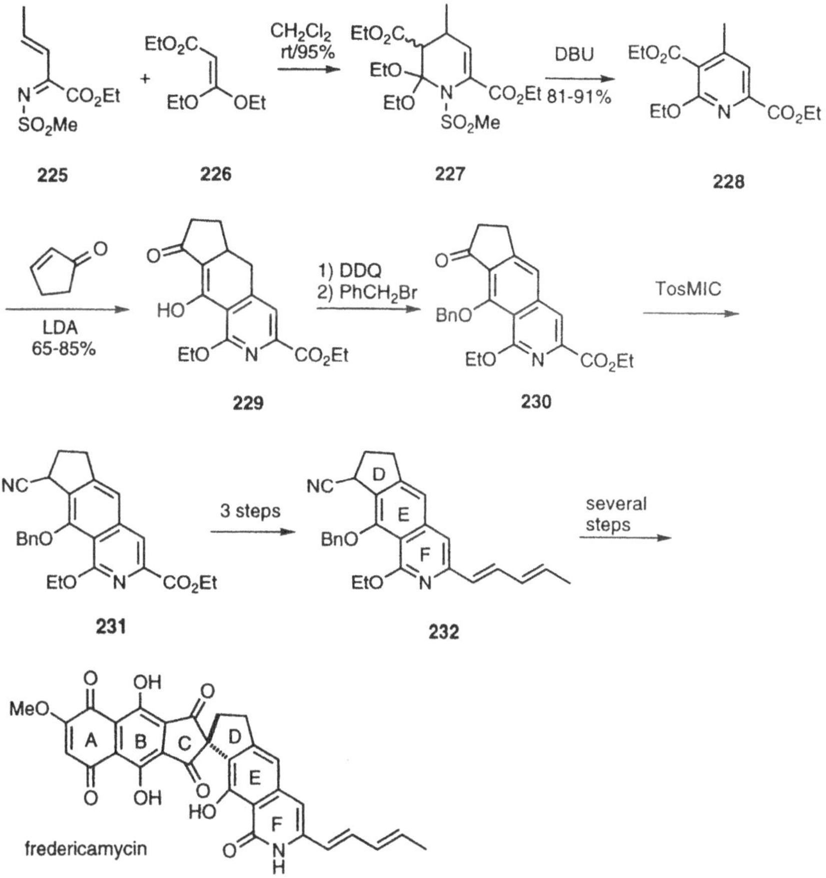

Scheme 40

6
[2+2]-Cycloadditions

A Russian group has described two types of [2+2]-cycloadditions of *N*-sulfonyl imines. It was found that both ketene and trimethylsilyl ketene react with chloral-derived *N*-sulfonyl imine **233** to afford β-lactams **234** in excellent yields [81] [Eq. (57)]. Although the compound produced from trimethylsilyl ketene appears to be a single isomer, the stereochemistry was not elucidated.

It was also reported that various alkoxy acetylenes undergo [2+2]-cycloadditions with chloral-derived *N*-sulfonyl imines to yield adducts **235** [82] [Eq. (58)]. Yields were claimed to be uniformly high.

R=H, SiMe$_3$ **233**   **234**   (57)

R=SiMe$_3$, SiEt$_3$, GeMe$_3$, Et
R'=Me, Et
R"=Me, Pr   **235**   (58)

7
Ene Reactions

In the first published example of an *N*-sulfonyl imino ene reaction, Achmatowicz and Pietraszkiewicz treated a variety of alkenes **236** with the *N*-tosyl imine derived from *n*-butyl glyoxylate (Scheme 41)[83, 84]. The products formed from these thermally promoted or ferric chloride-catalyzed transformations were γ,δ-unsaturated-α-amino acid derivatives **237**. However, no information was provided in the initial communications regarding the diastereoselectivity, if any, of this methodology. Weinreb and coworkers subsequently reinvestigated portions of the original work in order to elucidate the salient stereochemical features of these imino ene reactions and found that these transformations do in general show a high degree of stereospecificity [85].

When the *N*-tosyl imine prepared from ethyl glyoxylate was treated with *E*-2-butene at 150 °C, a 9:1 mixture of adducts **239** and **241** was produced (Scheme 42). It was rationalized that the reaction in fact proceeds via a concerted pericyclic mechanism [86] and formation of the major isomer **239** involves an *endo* ene transition state **238**, while the minor product **241** is formed from the

236   **237**

R	R'	R"	Yield (%)
H	H	H	70
Me	H	H	75
H	Me	H	79
H	H	Me	78
Me	H	Me	91
H	Ph	H	92
H	H	Ph	82

Scheme 41

238 (endo)   **239**

240 (exo)   **241**

Scheme 42

242 (endo)   **243**

244 (exo)   **245**

Scheme 43

exo transition state **240**. An attempt to apply this ene methodology to Z-2-butene under the same reaction conditions resulted in an 1:1 mixture of **239** and **241**. It was suggested that the loss of olefin stereochemical integrity here may result from thermal Z to E olefin isomerization during the reaction.

The ene reaction of cyclohexene was also investigated and it was found that using the ethyl glyoxylate-derived N-tosyl imine at 170°C led solely to sulfonamide ester stereoisomer **243** in good yield (Scheme 43). This result was rationalized by invoking a pericyclic imino ene transition state **242** in which the ester moiety prefers an *endo* orientation with respect to the alkene. Isomeric sulfonamide **245**, which would be derived from *exo* transition state **244**, was not produced in this reaction. The stereospecificity associated with the thermal reaction is again consistent with a concerted pericyclic ene process. Interestingly, when the reaction was run as originally described by Achmatowicz and Pietraszkiewicz [84] at 0°C using the Lewis acid catalyst FeCl₃, a mixture of stereoisomers **243** and **245** was obtained along with chlorinated products (cf. Scheme 17). It was suggested that the presence of the Lewis acid appears to change the reaction mechanism from a pericyclic one to an ionic stepwise Mannich-type process involving the olefin as outlined in Sect. 4.4.

It was also discovered that these reactions can be conducted intramolecularly and that they remain stereospecific. For example, imino ene reaction of the N-sulfonyl imine derived from **246** produced two *cis* δ-lactones **248** and **250** as a 9:1 mixture (Scheme 44). The formation of the major (E)-product **248** can be rationalized by invoking the more favorable pericyclic *endo* ene transition state **247**. The minor (Z)-product **250** would arise from *endo* ene transition state **249**, which suffers from $A^{1,3}$ strain between the allylic methyl substituent and the *cis* vinyl hydrogen. Rather surprisingly, the Z -olefin isomer corresponding to **246** gave the same 9:1 mixture as did the E isomer. Based on the related loss of stereoselectivity with Z-2-butene (vide supra) it was again postulated that the

Scheme 44

Z-alkene may be isomerizing to the *E* isomer prior to the imino ene cyclization. It was also reported that a γ-lactone can be prepared stereospecifically by the imino ene procedure. Thus, thermolysis of **251** gave exclusively the *cis* product **253**, presumably via *N*-tosyl aldimine **252** [Eq. (59)].

Kresze and coworkers have found about a three order of magnitude rate increase in imino ene reactions when an *N*-tosyl group is replaced with an *N*-perfluoroalkanesulfonyl moiety [87]. Thus, with the glyoxylate ester-derived imine and the one from chloral **255** as the enophiles, reactions with acyclic olefins are very rapid and occur at room temperature [Eq. (60)]. In these ene reactions one stereoisomeric homoallylic amine **256** was typically generated, although the stereochemistry was not elucidated. However, in one reported case a 1:1 mixture of diastereomers was obtained. Mechanistic studies based upon kinetic isotope effects seem to indicate that this reaction may in fact be a two-step process rather than a concerted one [87b].

251 → (*o*-DCB, Δ, 40%, −H₂O) → [**252**] → **253** (59)

254 + **255** R=CO₂Bu, CCl₃ → (CHCl₃, rt, 49-90%) → **256** (60)

Mikami et al. recently reported that *N*-sulfonyl imine **257**, derived from (–)-8-phenylmenthol glyoxylate, exhibited high diastereofacial selectivity in the imino ene reaction with methylenecyclohexane to afford homochiral α-amino ester derivatives **259** and **260** (96% de) in 60% yield [88] (Scheme 45). Presumably the diastereoselectivity is a consequence of a *syn*-chelated iminium complex **258** undergoing ene reaction with the olefin from the least sterically congested *si*-face. It might be noted that the imino ene reaction with isobutene and related chiral glyoxylate *N*-benzyl imine derivatives also gave similar diastereoselectivities, while reactions with imines derived from (*R*)- or (*S*)-phenylethylamine and (*S*)-α-amino esters were significantly less selective.

During a recent approach to the synthesis of the antibiotic polyoxins, methyl allene was treated with the *N*-sulfonyl imine from ethyl glyoxylate at 130 °C [89] [Eq. (61)]. However, the desired [2+2]-cycloadducts **262**–**264** were obtained in only very small amounts, while imino ene product **261** was the major product, formed in 33% yield.

Scheme 45

261 (33%)

262 (3%)   **263** (3%)   **264** (3%)   (61)

The only example of an *N*-sulfonyl ketimine participating in an ene reaction involves the tosyl imine of trifluoroacetone **266** [90] [Eq.(62)]. When heated with a terminal olefin such as allyl benzene (**265**) in refluxing xylene, imine **266** leads to ene product **267** in moderate yields. However, internal alkenes gave significantly lower yields of ene products. The inefficiency of the ene process with more highly substituted olefins was ascribed to unfavorable steric effects due to the bulky trifluoromethyl groups. Interestingly, with β-methyl styrene and allyl thiophenyl ether, [2+2]-cycloadducts were detected rather than ene products.

265   **266**   **267**   (62)

8
Miscellaneous Reactions

8.1
Metallations

Davis's group has looked at the metallation reactions of several types of *N*-sulfonyl imines in order to produce new oxaziridine reagents (vide infra). For example, cyclic sulfonyl imine **268** could be converted to a mono anion using LDA, but this species could not be successfully alkylated [91] [Eq. (63)]. However, a dianion **269** formed from **268** did C-alkylate with benzyl bromide and ethylene oxide, and gave *endo/exo* mixtures of products **270** and **271**.

$$R=CH_2Ph \qquad 1:1$$
$$R=CH_2CH_2OH \qquad 7:3$$

Davis et al. have extended this type of bis-metallation chemistry to a new method for synthesizing α-functionalized primary sulfonamides [5a]. *N*-Sulfonyl imines, such as **272,** are readily prepared from camphor and the corresponding sulfonamide using TiCl$_4$ as catalyst. These imines can be converted to the dianions **273** and monoalkylated to afford products **274** (Scheme 46). Hydrolysis of the alkylated imine affords a primary sulfonamide **275** in good overall yields. Unfor-

272 R=Me, Ph

273   E=RI, PhCHO, Et$_2$CO
ethylene oxide
PhCH$_2$Br

274

275 (ee < 35%)

Scheme 46

tunately, the diastereoselectivity observed in conversion of dianion **273** to products **274** was low, leading to low ees in the final products **275**.

It has also been demonstrated that it is possible to metalate camphor-derived *N*-tosyl imine **276** and to dichlorinate it to produce **277** [92] [Eq. (64)].

$$
\textbf{276} \quad \xrightarrow[\substack{\text{NCS} \\ 80\%}]{\substack{3\ \text{eq} \\ \text{NaHMDS}}} \quad \textbf{277} \tag{64}
$$

8.2
Oxidation

One can oxidize *N*-sulfonyl imines to the corresponding oxaziridines using oxidants such as a peracid or oxone [5b, 93, 94]. Thus, Davis and coworkers have developed a biphasic procedure for converting imine **278** to *trans*-oxaziridine **279**, in high yield [5b] [Eq. (65)]. Similarly, camphor-derived *N*-sulfonyl imines **277** and **268** can be oxidized to the *endo* oxaziridines **280** and **281**, respectively [92, 93] [Eq. (66), (67)]. Davis and others have now demonstrated the exceptional utility of oxaziridines as oxidants in organic synthesis and, in particular, the value of camphor-derived reagents, such as **280** and **281**, in asymmetric synthesis [94, 95].

$$
\textbf{278} \quad \xrightarrow[\substack{\text{NaHCO}_3/\text{H}_2\text{O} \\ \text{BnNEt}_3\text{Cl}/\text{CHCl}_3 \\ 88\%}]{\text{MCPBA}} \quad \textbf{279} \tag{65}
$$

$$
\textbf{277} \quad \xrightarrow[\substack{\text{K}_2\text{CO}_3/\text{CH}_2\text{Cl}_2 \\ \text{Aliquat 336} \\ 90\text{-}95\%}]{\text{AcOOH}} \quad \textbf{280} \tag{66}
$$

$$
\textbf{268} \quad \xrightarrow[\substack{\text{K}_2\text{CO}_3/\text{H}_2\text{O} \\ 84\%}]{\text{oxone}} \quad \textbf{281} \tag{67}
$$

8.3
Eliminations

Glass and Hoy have reported that treating an *N*-tosyl imine derived from an aromatic aldehyde with cyanide in HMPT leads to aryl nitriles in generally good yields [96]. It was proposed that this transformation might involve intermediates shown in Eq. (68). Alternatively, a dianion corresponding to the mono anions shown could be involved here.

$$\text{(68)}$$

9
Perspectives

It seems clear from the chemistry outlined in this review that *N*-sulfonyl imines have significant future potential as synthons in organic synthesis. The recently improved access to this class of compounds has been the primary contributor to their increased use compared with one or two decades ago. However, improvements in old methods and discovery of new ones are still required for generation of *N*-sulfonyl aldimines and, in particular, *N*-sulfonyl ketimines. It would also be useful to have more structural information on these species, including data on geometrical isomerization. Since many *N*-sulfonyl imines are quite stable, NMR and X-ray studies would seem to be quite feasible. Finally, inventive organic chemists should consider using these synthons when investigating any type of chemistry involving electron-deficient imines.

10
Addendum

Stabilized Carbanions and Enol Derivatives A recent report has described the *erythro*-selective aldol reaction of *N*-tosyl aldimines from aromatic and conjugated aldehydes with methyl isocyanoacetate catalyzed by a gold complex [97]. For example, condensation of the *N*-tosyl imine of benzaldehyde with methyl isocyanoacetate in the presence of an Au (I) catalyst gave a very high yield of the imidazolines **282** and **283** in an 11:89 ratio (Scheme 47). The major *cis* product could be hydrolyzed to afford the *erythro* vicinal damine **284**. Other metal catalysts such as CuCl, AgOTf, PdCl$_2$(MeCN)$_2$, and [RhCl(COD)]$_2$ gave lower *cis* selectivities in the initial step. In addition, *N*-phenyl, *N*-*p*-carbomethoxyphenyl and *N*-diphenylphosphinyl imines were not reactive in this process. Interestingly, the corresponding reactions with aldehydes have previously been reported by this group to yield primarily the *trans* disubstituted oxazolines, although no rationale for the difference in selectivity between sulfonyl imines and aldehydes was provided.

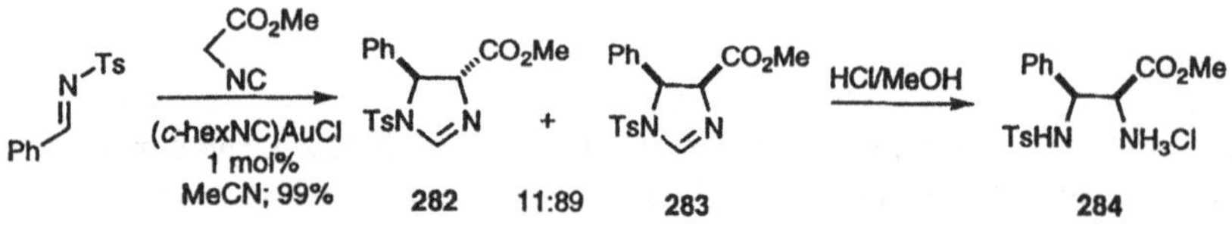

Scheme 47

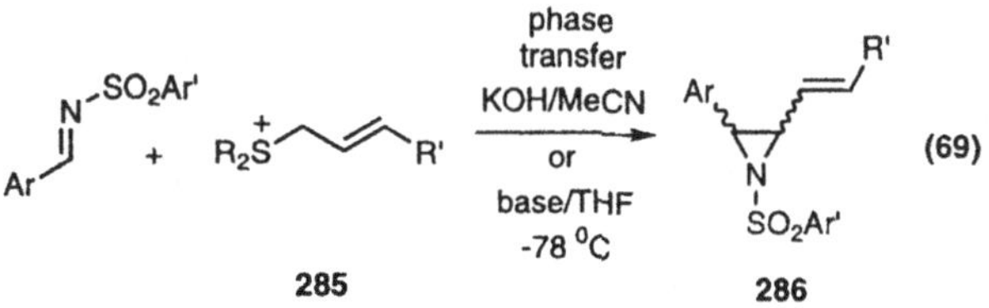

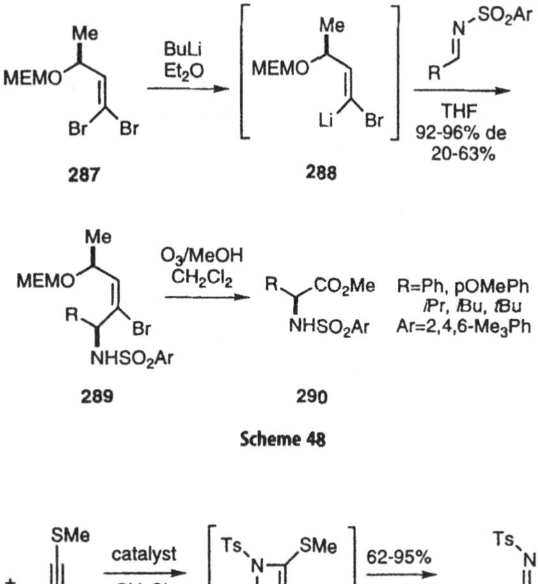

(69)

Scheme 48

(70)

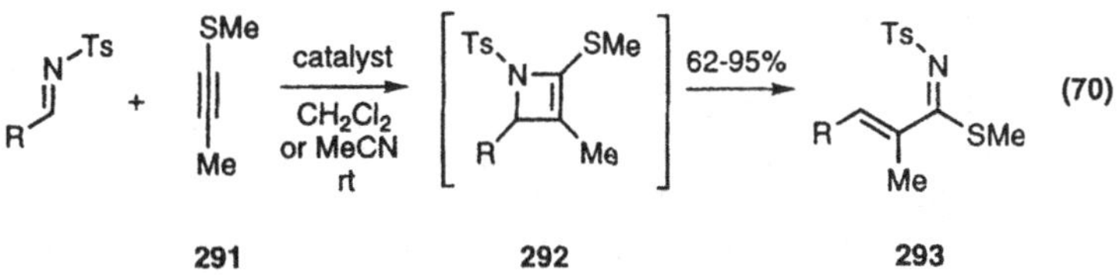

Dai and coworkers have investigated the addition of ylides derived from allylic sulfonium salts to *N*-sulfonyl imines derived from aromatic aldehydes to produce 2-vinyl aziridines [98]. The general reaction studied is shown in Eq. (69), where a sulfonium salt **285** can be converted to its ylide *in situ* under either phase transfer conditions, or by use of a strong amides base, and then combined with an *N*-sulfonyl imine to provide a 2-vinyl aziridine **286**. In general, stereoselectivity was not high and mixtures of *cis* and *trans* aziridines were formed in all cases. Slight variations in the ratio of isomers occurred depending upon the substitution on the imine and ylid components. Other types of *N*-substituted imines did not afford any aziridines.

An enantioselective route to α-amino acid derivatives has been developed which utilizes addition of a scalemic vinyllithium species to an *N*-sulfonyl imine [99]. Metallation of dibromide **287**, derived from enantiomerically pure lactic acid, is regioselective and affords lithiated species **288** which adds to a variety of *N*-mesitylsulfonyl imines to give adducts **289** in diastereomeric excesses above 90% (Scheme 48). Ozonolysis of the olefinic moity of purified adducts **289** then afforded enantiomerically pure α-amino acid derivatives **290**.

[2+2]-Cycloadditions. It was found recently that *N*-tosyl imines of a variety of aldehydes react with alkynyl sulfides in the presence of a Lewis acid catalyst to afford α, β- unsaturated thioimidates [100]. Thus alkynyl sulfide **291** combines with an *N*-sulfonyl imine in the presence of a catalyst such as BF_3 etherate, $Yb(OTf)_3$, $Sc(OTf)_3$ or $Ln(OTf)_3$ to give an imidate **293** [Eq. (70)]. It is believed that this transformation occurs through an initial [2+2]-cycloaddition of the reactants to form an azetine **292**. Other types of *N*-substituted imines were found to react with alkynyl sulfides under these conditions to provide different sorts of products and not α, β-unsaturated imidates like **293**.

References

1. For some reviews of the chemistry of *N*-acyl imines, see a) Zaug, HE (1982) Synthesis 85, 181 b) Hiemstra H, Speckamp WN (1991) Additions to *N*-acyl iminium ions. In: Trost BM, Fleming I (eds) Comprehensive Organic Synthesis, Pergamon, Oxford, Vol 2, p 1047 c) Scola PM, Weinreb SM (1989) Chem Rev 89:1525 d) Boger DL, Weinreb SM (1987) Hetero Diels-Alder methodology in organic synthesis, Academic Press, Orlando, Chap 2 e) Malassa I, Matthies D (1987) Chem-Ztg 111:181, 253
2. Lichtenberger J, Fleury J-P, Baretta B (1955) Bull Soc Chim Fr 669
3. Kretow AJ, Abrazhanova EA (1957) J Gen Chem USSR (Engl Trans) 27:1993
4. a) Jennings WB, Lovely CJ (1988) Tetrahedron Lett 29:3725 b) Jennings WB, Lovely CJ (1991) Tetrahedron 47:5561
5. a) Davis FA, Zhou P, Lal GS (1990) Tetrahedron Lett 31:1653 b) Vishwakarma LC, Stringer OD, Davis FA (1987) Org Synth 66:203
6. Albrecht R, Kresze G, Mlakar B (1964) Chem Ber 97:483
7. Albrecht R, Kresze G (1965) ibid 98:1431
8. For a review of *N*-sulfinyl compounds, see Bussas R, Kresze G, Munsterer H, Schwobel A (1983) Sulfur Rep 2:215
9. Hori T, Singer SP, Sharpless KB (1978) J Org Chem 43:1456
10. A report exists of the isolation of heterocycles like 2; Pozdnyakova TM, Sergeyev NM, Gorodetskaya NI, Zefirov NS (1972) Int J Sulfur Chem 2:109
11. Melnick MJ, Freyer AJ, Weinreb SM (1988) Tetrahedron Lett 29:3891
12. Sisko J, Weinreb SM (1989) Tetrahedron Lett 30:3037
13. Sisko J, Weinreb SM (1990) J Org Chem 55:393 see also Hegedus LS, Holden MS (1986) ibid 51:1171
14. Alexander MD, Anderson RE, Sisko J, Weinreb SM (1990) J Org Chem 55:2563 see also Hegedus LS, McKearin JM (1982) J Am Chem Soc 104:2444
15. Ralbovsky, JL, Kinsella MA, Sisko J, Weinreb SM (1990) Synth Commun 20:573
16. Zhu, S-Z, Chen Q-Y (1991) J Chem Soc Chem Commun 732

17. Trost BM, Marrs C (1991) J Org Chem 56:6468
18. Brown C, Hudson RF, Record KAF (1978) J Chem Soc, Perkin Trans 2 822 and references cited therein
19. a) Boger DL, Kasper AM (1989) J Am Chem Soc 111:1517 b) Boger DL, Corbett WL, Curran TT, Kasper AM (1991) J Am Chem Soc 113:1713
20. Boger DL, Corbett WL (1992) J Org Chem 57:4777
21. Barton DHR, Jaszberenyi JC, Theodorakis EA (1991) Tetrahedron 47:9167
22. Pickard PL, Tolbert TL (1961) J Org Chem 26:4886
23. Georg GI, Harriman GCB, Peterson SA (1995) J Org Chem 60:7366
24. Hart DJ, Kanai K, Thomas DG, Yang T-K (1983) J Org Chem 48:289
25. a) Shono T, Matsumura Y, Katoh S, Takeuchi K, Sasaki K, Kamada T, Shimizu R (1990) J Am Chem Soc 112:2368 b) Shono T, Matsumura Y, Uchida K, Nakatani F (1988) Bull Soc Chem Jpn 61:3029 c) Ross, SD, Finkelstein M, Rudd EJ (1972) J Org Chem 37:2387
26. Han G, Weinreb SM , unpublished results
27. Pines SH, Purick RM, Reamer RA, Gal G (1978) J Org Chem 43:1337
28. cf Han G, McIntosh MC, Weinreb SM (1994) Tetrahedron Lett 35:5813 and references cited therein
29. Ahman J, Somfai P (1992) Tetrahedron 48:9537
30. Davis FA, Kluger EW (1976) J Am Chem Soc 98:302
31. Prosyanik AV, Kol'tsov NY, Romanchenko VA, Belov VV, Burmistrov KS, Loban SV (1987) J Gen Chem USSR (Engl Trans) 23:335
32. For X-ray studies on the bis-N-tosyl imine of p-benzoquinone, see Shuets AE, Mishnev AF, Vleidelis YY (1978) Zh Strukt Khim 19:544
33. For ^{19}F NMR studies and some chemistry of perfluorinated N-sulfonyl imines, see Petrov VA, Mlsna TE, Desmarteau DD (1994) J Fluorine Chem 68:277
34. (a) Ponzo VL, Kaufman TS (1995) Synlett 1149 (b) Kaufman TS (1996) J Chem Soc Perkin Trans I 2997
35. Reuter M, German patent 847,006 (Chem Abstr (1956) 50:2669)
36. Nemoto H, Kubota Y, Yamamoto Y (1990) J Org Chem 55:4515 see also Lin Y-R, Zhou X-T, Dai L-X, Sun J (1997) J Org Chem 62:1799, Kai H, Iwamoto K, Chatani N, Murai S (1996) J Am Chem Soc 118:7634
37. Shimada S, Saigo K, Abe M, Sudo A, Hasegawa M (1992) Chemistry Lett 1445
38. Weinreb SM, Sisko J, unpublished results
39. cf Krow GR, Pyun C, Marakowski J (1974) J Org Chem 39:2449
40. Kobayashi S, Araki M, Ishitani H, Nagayama S, Hachiya I (1995) Synlett 233
41. Saigo K, Shimada S, Hasegawa M (1990) Chemistry Lett 905
42. Ralbovsky JL, Scola PM, Sugino E, Burgos-Garcis C, Weinreb SM, Parvez M (1996) Heterocycles 43:1497
43. Somfai P, Ahman J (1992) Tetrahedron Lett 33:3791
44. a) Lu Z-H, Zhou W-S (1993) J Chem Soc Perkin Trans 1 393 b) Lu Z-H, Zhou W-S (1993) Tetrahedron 49:4659
45. Zhou, W-S, Lu Z-H, Wang Z-M (1991) Tetrahedron Lett 32:1467
46. Sisko J, Henry JR, Weinreb SM (1993) J Org Chem 58:4945
47. cf Takano S, Iwabuchi Y, Ogasawara K (1987) J Am Chem Soc 109:5523
48. Overman LE, Blumenkopf TA (1986) Chem Rev 86:857
49. Fleming I, Dunogues J, Smithers R (1989) Org React 37:57
50. McIntosh MC, Weinreb SM (1993) J Org Chem 58:4823
51. Garigipati RS, Tschaen DM, Weinreb SM (1990) J Am Chem Soc 112:3475. Weinreb SM (1995) in Studies in natural products chemistry (Rahman A, ed) Elsevier, Amsterdam, Vol 16 (Part J) p 3
52. see also Magnus P, Lacour J, Coldham I, Mugrage B, Bauta WB (1995) Tetrahedron 51:11087
53. Volkmann RA (1991) Nucleophilic addition to imines and imine derivatives in Comprehensive organic synthesis (Trost BM, Fleming I, eds) Pergamon, Oxford, Vol 1, p. 355

54. Reetz MT, Jaeger R, Drewlies R, Hubel M (1991) Angew Chem Int Ed Engl 30:103 see also Hopman JCP, van den Berg E, Ollero Ollero L, Hiemstra H, Speckamp WN (1995) Tetrahedron Lett 36:4315

55. a) Cherkauskas JP, Borzilleri RM, Sisko J, Weinreb SM (1995) Synlett 527 b) Cherkauskas JP, Klos AM, Borzilleri RM, Sisko J, Weinreb SM, Parvez M (1996) Tetrahedron 52:3135

56. Zaugg HE (1965) Org React 14:52 see also Negash K, Nichols DE (1996) Tetrahedron Lett 37:6971. Ishizaki M, Hoshino O, Iitaka Y (1992) J Org Chem 57:7285

57. Il' in GF, Kolomiets AF, Sokol'skii GA (1979) J Org Chem USSR (Engl Trans) 15:2012

58. Hamley P, Holmes AB, Kee A, Ladduwahetty T, Smith DF (1991) Synlett 29

59. Monti L, Felici L (1940) Gazz Chim Ital 70:375

60. Weinreb SM (1991) Heterodienophile additions to dienes in Comprehensive organic synthesis (Trost BM, Fleming I, eds) Pergamon, Oxford, Vol 5, p 401

61. Kresze G, Albrecht R (1964) Chem Ber 97:490

62. Kresze G, Wagner U (1972) Liebig's Ann Chem 762:106

63. a) Rijsenbrij PPM, Loven R, Wijnberg JBPA, Speckamp WN, Huisman HO (1972) Tetrahedron Lett 1425 b) Loven RP, Zunnebeld WA, Speckamp WN (1975) Tetrahedron 31:1723

64. Kasper F, Dathe S (1985) J Prakt Chem 327:1041

65. a) Krow G, Rodebaugh R, Marakowski J (1973) Tetrahedron Lett 1899 b) Krow GR, Dyun C, Rodebaugh R, Marakowski J (1974) Tetrahedron 30:2977

66. see also Zunnebeld WA, Speckamp WN (1975) Tetrahedron 31:1717. In this example, poorer regioselectivity was observed

67. a) Barco A, Benetti S, Baraldi PG, Moroder F, Pollini GP, Simoni D (1982) Liebig's Ann Chem 960 b) Maggini M, Prato M, Scorrano G (1990) Tetrahedron Lett 31:6243

68. Holmes AB, Raithby PR, Thompson J, Baxter AJG, Dixon J (1983) J Chem Soc, Chem Commun 1491. Holmes AB, Kee A, Ladduwahetty T, Smith DF (1990) J Chem Soc, Chem Commun 1412

69. Heintzelman GR, Weinreb SM, Parvez M (1996) J Org Chem 61:4594

70. a) Hamada T, Sato H, Hikota M, Yonemitsu O (1989) Tetrahedron Lett 30:6405 b) Hamada T, Zenkoh T, Sato H, Yonemitsu O (1991) ibid 32:1649

71. Tripathy R, Franck RW, Onan KD (1988) J Am Chem Soc 110:3257

72. a) Holmes AB, Thompson J, Baxter AJG, Dixon J (1985) J Chem Soc Chem Commun 37 b) Birkinshaw TN, Tabor AB, Holmes AB, Kaye P, Mayne PM (1988) J Chem Soc Chem Commun 1599 c) Birkinshaw TN, Tabor AB, Holmes AB, Raithby PR (1988) J Chem Soc Chem Commun 1602

73. Hamley P, Helmchen G, Holmes AB, Marshall DR, MacKinnon JWM, Smith DF, Ziller JW (1992) J Chem Soc Chem Commun 786

74. Poll T, Metter JO, Helmchen G (1985) Angew Chem Int Ed Engl 24:112

75. (a) McFarlane AK, Thomas G, Whiting A (1993) Tetrahedron Lett 34:2379 (b) McFarlane AK, Thomas G, Whiting A (1995) J Chem Soc Perkin Trans 1:2803

76. (a) Abramovitch RA, Stowers JR (1984) Heterocycles 22:671 (b) Abramovitch RA, Shinkai I, Mavunkel BJ, More KM, O'Connor S, Ooi GH, Pennington WT, Srinivasan PC, Stowers JR (1996) Tetrahedron 52:3339

77. Fujii T, Kimura T, Furukawa N (1995) Tetrahedron Lett 36:1075

78. a) Boger DL, Kasper AM (1989) J Am Chem Soc 111:1517 b) Boger DL, Corbett WL, Wiggins JM (1990) J Org Chem 55:2999 c) Boger DL, Curran TT (1990) J Org Chem 55:5439 d) Boger DL, Corbett WL, Curran TT, Kasper AM (1991) J Am Chem Soc 113:1713

79. see also Orsini F, Sala G (1989) Tetrahedron 45:6531

80. a) Boger DL, Zhang M (1992) J Org Chem 57:3974 b) Boger DL, Huter O, Mbiya K, Zhang M (1995) J Am Chem Soc 117:11839

81. Novikova OP, Livantova LI, Zaitseva GS (1990) Zh Obsch Khim 59:2630

82. Zaitseva GS, Novikova OP, Livantsova LI, Petrosyan VS, Baukov YI (1992) Zh Obsch Khim 61:1389 see also Sriraj V, Deshmukh ARAS, Puranik VG, Bhawal BM (1996) Tetrahedron: Asymmetry 7:2735

83. For a review of imino ene reactions see: Borzilleri RM, Weinreb SM (1995) Synthesis 347

84. a) Achmatowicz O, Pietraszkiewicz M (1976) J Chem Soc Chem Commun 484 b) Achmatowicz O, Pietraszkiewicz M (1981) J Chem Soc Perkin Trans 1:2680

85. a) Tschaen DM, Weinreb SM (1982) Tetrahedron Lett 23:3015 b) Tschaen DM, Turos E, Weinreb SM (1984) J Org Chem 49:5058
86. For a recent theoretical treatment of imino ene reactions, see Thomas BE, Houk KN (1993) J Am Chem Soc 115:790
87. a) Braxmeier H, Kresze G (1985) Synthesis 683 b) Starflinger W, Kresze G, Huss K (1986) J Org Chem 51:37
88. Mikami K, Kaneko M, Yajima T (1993) Tetrahedron Lett 34:4841
89. Baumann H, Dulhaler RO (1988) Helv Chim Acta 71:1025
90. Shimada T, Ando A, Takagi T, Koyama M, Miki T, Kumadaki I (1992) Chem Pharm Bull 40:1665
91. Davis FA, Weismiller MC, Lal GS, Chen BC, Przeslawski RM (1989) Tetrahedron Lett 30:1613
92. Davis FA, Thimma Reddy R, Weismiller MC (1989) J Am Chem Soc 111:5964
93. a) Chen BC, Murphy CK, Kumar A, Thimma Reddy R, Zhou P, Lewis BM, Gala D, Mergelsberg I, Scherer D, Buckley J, Dibenedetto D, Davis FA (1995) Org Synth 73:159 b) Towson JC, Weismiller MC, Lal GS, Sheppard AC, Davis FA, Org Synth 69:158
94. Davis FA, Thimma Reddy R, Han W, Reddy RE (1993) Pure Appl Chem 65:633
95. Davis FA, Sheppard AC (1989) Tetrahedron 45:5703
96. Glass RS, Hoy RC (1976) Tetrahedron Lett 1777 and 1781
97. Hayashi T, Kishi E, Soloshonok VA, Uozumi Y (1996) ibid 37:4969
98. (a) Li A-H, Dai L-X, Hou X-L (1996) J Chem Soc, Chem Commun 491 (b) Li A-H, Dai L-X, Hou X-L (1996) J Chem Soc Perkin Trans 1 867 (c) Li A-H, Dai L-X Hou X-L, Chen M-B (1996) J Org Chem 61:4641 (d) Li A-H, Dai L-X, Hou X-L (1996) J Chem Soc Perkin Trans I 2725
99. Braun M, Opdenbusch K (1993) Angew Chem Int Ed Engl 32:578
100. Ishitani H, Nagayama S, Kobayshi S (1996) J Org Chem 61:1902
101. For new reactions of *N*-sulfonyl imines see (a) Aggarwal VK, Thompson A, Jonez RVH, Standen MCH (1996) J Org Chem 61:8368 (b) Charette A, Giroux A (1996) Tetrahedron Lett 37:6669

Springer
and the
environment

At Springer we firmly believe that an international science publisher has a special obligation to the environment, and our corporate policies consistently reflect this conviction.
We also expect our business partners – paper mills, printers, packaging manufacturers, etc. – to commit themselves to using materials and production processes that do not harm the environment.

Made in the USA
Monee, IL
07 July 2026

56552002R00109